the dairy

the dairy

LEANNE KITCHEN

MURDOCH BOOKS

contents

dairy products

Since time immemorial, humans have been consuming the products and parts of animals, and among these, milk must surely rank as one of the most important. Throughout much of history, milk, collected daily from common farmyard animals, provided vital nourishment — including valuable protein — for many communities too poor to afford much, if any, meat.

Animals were first domesticated some 12,000 years ago, so our relationship with sheep, cows and goats goes back a very long way indeed. These days, we invest little to no effort at all in our procurement of milk, cream, cheese and the many other products we loosely term as 'dairy'. A trip to the nearest corner store or supermarket with a bit of spare change easily gets us a carton of milk, some sour cream or buttermilk, a block of cheese or a tub of ricotta. Most sizable towns and certainly any city will have a specialist deli or a dedicated cheese shop where less mainstream cheesy offerings can be purchased, cut fresh from a wedge, slab or square.

Over the centuries, initially for reasons to do with preserving fresh milk (which is highly perishable), a whole raft of milk-based products have been devised, from the ancient (such as yoghurt and butter) to the modern (such as cream cheese and condensed milk). There is now an enormous range of dairy foods at our disposal, and such foods occur widely in almost every food culture in the world; China is a notable exception, although even this situation is rapidly changing.

As a topic, 'The Dairy' is an enormous one, and it is beyond the scope of this book to examine the myriad, and fascinating, forms in which dairy products appear around the globe. Across the Indian subcontinent, for example, paneer, a fresh white cheese used in everything from sweets to vegetarian curries, is a staple. In Tibet, life would be unthinkable without yak butter, used most famously in their tea, and throughout Greece and the Middle East, yoghurt has long been fermented with grain then sun dried and powdered, to provide instant nourishment throughout the winter months.

Milk is a miraculous food. All mammals produce it, but that for human consumption (apart from breast milk for newborns, of course) comes from ruminants. Milk is designed by nature as a sustaining food for infant mammals, and somewhere along the way, humans worked out that it was a highly versatile foodstuff. Originally animals were kept as working beasts, and their meat and pelts used once they died, but with the advent of dairying, animals could be utilised as a food source while still alive. The principal animals exploited for their milk are cows, sheep, goats, water buffalo (common throughout Asia, parts of the Middle East and in Southern Italy), yak (found in Tibet and Central Asia) and camels (used throughout Northern Africa). Ruminants such as these have a complex digestion that can efficiently process large amounts of feed into copious quantities of milk. The great thing about such animals is that they pose no competition for humans as far as their diet goes, making for a harmonious co-existence.

Until the Industrial Revolution, all dairying was done on a small scale, on farms. People produced as much milk, butter, cheese or yoghurt as they needed, with perhaps a little left over to sell or barter. Supply to urban areas was tricky before refrigeration and it was not uncommon for people (mainly children) to die from drinking contaminated milk. With the advent of the railway in the early to mid-nineteenth century, milk could be regularly and freshly supplied, and improvements in hygiene meant that milk was more reliably 'clean'. Milking machines were invented, and animals (particularly cows) no longer served a dual purpose as both workers and milkers. Slowly, a dedicated dairy industry developed. Louis Pasteur, the French scientist who pioneered the process of heat-treating milk, also pioneered the standardisation of cultures and bacteria used in cheese making.

Now, in the twenty-first century, dairying is (in the developed world at least) far removed from the traditional image of milkmaids hand-collecting milk from individually named cows into wooden pails. Herds are enormous; cows are bred for their milk-producing efficiency and the entire process of extracting milk and providing it, and other dairy foods, to consumers has become clinical in the extreme, with dairy products now extremely mass produced. As with any food, it is worthwhile to seek out products made in small batches according to artisanal methods, and tasting the difference. What we have gained in the cheapness of our dairy foods and in the efficiency of their production, we have lost in the richness that comes when natural, slower processes allow for slight and interesting variations in taste and texture. These days there is a growing interest in raw milk and in products made using it; many traditional European cheeses are still made using raw milk and their fans claim them to be far superior in flavour and texture to equivalents made using pasteurised milk. In most industrialised countries, it is difficult to procure raw milk; it can be a completely safe food but needs to come from suppliers who run a scrupulously clean dairy.

Of all dairy foods, perhaps none provides so much fascination as cheese. The sheer number of cheeses is enormous. The potential for new ones is almost as vast — unique varieties are constantly being developed in specialist dairies around the world. Cheese dates back to ancient eras; by the time of the Greeks and Romans, it was an important trade commodity and was mentioned widely in the literature of the day (including in Homer's *Odyssey*, in which the Cyclops Polyphemus is described making it in his cave). The Roman writer Apicius lists the steps in cheese making when writing in the first century AD. The Roman word for cheese was *caseus*, from which the word casein (the milk protein that is curdled to make cheese) is derived. The French and Italian words for cheese, *fromage* and *formaggio* respectively, come from the Greek word for form or mould — *formos*.

When all is said and done, cheese is simply a way to preserve milk for later use, and that is why it was originally made. Essentially, cheese is made by curdling fermented milk then draining off the liquid that forms (the whey). The curds are generally curdled using rennet, a substance extracted from the stomach of suckling animals, and flavoured with salt. It isn't known when rennet was first used, but there is speculation it was by the Etruscans, as early as the fifth or sixth century BC. It was the Romans who pioneered the technique of cooking curds in cheese making, allowing for longer-keeping, harder cheeses that could easily be transported and traded. The combination of bacteria used to ferment the milk, in combination with the type of milk and the particular type of cheese-making process employed, plus the action of time as a cheese is matured, are the variables that give rise to the large number of cheeses we know today.

This cheese-making knowledge accumulated slowly over centuries. Initially, folk made one or two types of cheese in their particular locale, the character of which depended on the local conditions (such as bacteria present, climate, type of animal and pasture, and storage conditions).

Like a bread dough or a batch of fermenting wine, a cheese is a living thing — the bacteria that feed on the cheese produce enzymes that develop flavour, texture and aromas over time.

Cheese making was well established in Europe by the Roman era. During the Dark Ages, when the Roman Empire had all but disappeared, it truly took off, with monasteries and large farming estates becoming the centres of cheese making. In areas where large, hard, aged cheeses were made (these had better keeping qualities and were more transportable than fresh cheeses), cheese-making co-operatives sprang up early; large cheeses required more milk than single producers could supply. An example is gruyère, which was made by co-operatives as early as the thirteenth century.

Perhaps no other country has given the world such a diversity of cheese as has France; in 1962, Charles de Gaulle, the then French president, famously pondered how it was possible to govern a country that had 246 cheeses. France has certainly given us some of the most famous cheeses (including brie, camembert and roquefort), but many other nations have large repertoires of traditional cheeses. Portugal, Spain, Italy, England, Holland, the Balkan and Scandinavian countries, Greece, Russia, Mexico, Afghanistan, Syria and many others besides have their own cheeses.

Throughout history, cheese has been the food of sophisticates and the humble alike; fresh, simply made cheese use to be called 'white meat', such was its importance as a protein food on the tables of the poor. The wealthy could afford the complex cheeses that required skill and long aging to construct and ripen. By the turn of the last century, the art of cheese making was at its zenith; unfortunately the industrialised processes that then bought cheese fairly cheaply to more tables ultimately led to its standardisation. In 1973 in France, a benchmark called the Fromage Appellation d'Origine Contrôlée was introduced to protect the integrity of traditional cheeses and their methods of making; other countries, such as Italy, also have such labelling and protection systems in place. The vast majority of modern cheese, though, is made in factories.

However, across the world, there has been a resurgence of interest in both the making and the eating of good cheeses. Small, 'boutique'-scale cheese makers are increasing in number. Farmers' markets are also becoming popular and are often a good place to sample locally made cheeses from small producers. People are interested in the making (and consuming) of goat's cheese and sheep's milk cheeses too, and excellent examples of such cheeses can usually, with a little persistence, be found. Many cheeses are ideal for using in cooking but sometimes there is nothing better than simply savouring a beautifully made specimen, at the peak of ripeness and at room temperature with a few well-chosen accompaniments. In this book, we will explore ways to do just this, and provide delicious recipes as well as fascinating background and practical information too.

Perhaps the most significant dairy product, and certainly the most various, is cheese. It is simply a form of preserved milk, but that description does not do justice to the complexity of the cheese-making process, or the wonderful, versatile and very different products that result. The rest of this section examines this process.

HOW CHEESE IS MADE

All cheeses are made in essentially the same way: lactic acid bacteria convert the lactose (milk sugar) in milk into lactic acid. Rennet is then added, causing the proteins in the milk to curdle and resulting in the formation of curds and whey. Curds form due to both the rennet and the lactic bacteria, and the amount of each present affects the size and form of the curds. Rennet makes rubbery curds; lactic acid more fragile curds. Fresh cheeses are made from largely acid curds, and semi-hard and hard cheeses from rennet curds.

The whey is then drained off; this can be done in a variety of ways. The cheese can simply be hung in a perforated mould or bag and left to drain, which is how fresh cheeses are made. Some cheeses require the curds to be cut into many small pieces, creating a large surface area and the ability for a lot of moisture to drain out. These curds are then left to drain, or can be pressed to remove the liquid, depending upon the type of cheese to be made (firmer cheeses are made using small, cut curds). Some cheeses, notably the larger, hard ones, have their curd 'cooked' in the whey (by heating to 55°C/130°F) before draining. This forces water out of the curds and also causes particular, complex flavours to develop.

The curds are salted, either by adding dry salt or by soaking the formed but immature cheese in brine for a certain time. Salt adds flavour, prevents the formation of unwanted bacteria, and contributes to the ripening process by moderating the action of enzymes. Most cheese has salt; pecorino, roquefort and feta are among the saltiest, at about 5 per cent salt by weight. Of the traditional cheeses, emmental is one of the least salty. Once the cheese is shaped, it is allowed time to ripen, during which time — due to the actions of various microbes, moulds, 'good' bacteria and enzymes — it develops texture and flavour.

Although this all may sound terribly simple, making cheese is a highly skilled enterprise, requiring expertise and experience. A cheese is a living thing, complete with a life-cycle. This begins with the milk used; even within breeds, the flavour of milk can vary. Milk also differs according to the season, the pasture and feed available to the animal, and where the animal is in its lactation cycle (milk taken early in an animal's breeding cycle is richer and higher in fat and protein than that taken from the end). Milk extracted in the evening is richer in fat than morning milk — which is why many cheeses are made both using morning and evening milk while some, like the French reblochon, are made entirely from second (or evening) milkings. The unique qualities of the land where the cheese is made (the climate, minerals in the ground and the quality of the pasture, for example) give cheeses from certain locales unique characteristics, which is why some cheeses just don't taste the same when made elsewhere. Unless of the fresh type, a cheese develops from being fairly bland, uninteresting curds into a complex, character-filled finished product ready to eat; beyond this point, it moves beyond maturity and into a phase of collapse and decay. The cheese-making process demands know-how and time on the part of the maker. It also requires some knowledge on the part of the consumer (or at least the cheese merchant selling it to them) so that it can be enjoyed at its optimum best.

Some cheeses ripen in weeks, while others, such as parmesan, take a year or so to age before ready to eat. The aging (or 'ripening') process is always overseen by a skilled person who understands just the correct balance of humidity, temperature, time, enzymes and bacteria needed to achieve the best result possible. The best cheese shops will purchase slightly immature cheeses and oversee the final stages of ripening themselves so that they can offer them for sale at their peak of ripeness; the person who does this is called an *affineur*. Large, industrial-scale cheese producers partially ripen their cheeses, then stall further ripening by refrigeration. They ship them to retailers when under-ripe so the cheeses better withstand the rigours of transportation and storage until sale; if buying such cheese, always check the recommended use-by date to calculate how long it still needs to ripen. A specialist cheese retailer will be able to advise you on the condition of each cheese and recommend the ones best for eating now.

CLASSES OF CHEESE

With cheeses encompassing such an incredibly diverse range of ages, textures and flavours, how should one categorise them? There are a few methods and none is entirely satisfactory, as there are always cheeses that don't fit neatly into just one category; some cross over from one group to another as they age. For example, pecorino is quite yielding and soft when young but matures into a very hard grating cheese. The most common method is to class them by their moisture content, and thus cheeses are usually described as being 'soft' or 'fresh', 'semi-soft', 'semi-hard' or 'hard'.

The harder the cheese, the lower the moisture content and the longer its life; conversely, the softer the cheese, the higher the moisture content and the shorter the shelf life. Some other ways of categorising cheeses are by the method of making ('surface-ripened' or *'pasta filata'*, for example), by fat content ('double cream' or 'triple cream', for example) or by the type of milk used (sheep's or goat's milk cheese, for example). Don't be lulled, though, into thinking a hard cheese has a lower fat content than a triple cream cheese — the former claims a fat content of about 35 per cent, while the latter can have a hefty 75 per cent. The fat, however, is in the solid part of the cheese, and a soft cheese (such as a triple cream) contains a relatively high proportion of water, while a hard cheese has a low amount of water. For example, parmesan has about 35 per cent fat and 30 per cent water, which translates to 21 grams (³/₄ oz) of fat for every 100 g (3¹/₂ oz) of cheese. A brie contains about 49 per cent fat and 57 per cent water, so also delivers about 21 grams of fat per 100 g. Cheese loses moisture as it ages but the fat content remains the same, so the most accurate way for cheese makers to describe the fat content is in relation to its dry weight, not its total weight.

BUYING CHEESE

In his *Guide de Fromage*, Pierre Androuët insists that 'a cheesemonger worthy of the name will offer only cheeses that may be eaten within the next 48 hours. For no cheese … will benefit from being kept too long at home.' While we might not all be fortunate enough to have access to such a cheesemonger, Androuët's statement does underline how important it is to seek out a good, specialist cheese retailer; in general, you won't get great cheese from the supermarket or other mass-market outlets. A cheesemonger will offer cheeses cut to order from large wheels or blocks and most often you will be able to taste a little before purchasing. Plastic-wrapped, pre-cut pieces of cheese suffer from flavour deterioration with prolonged surface exposure to light, air and even to the plastic wrap. A specialist store will keep cheeses at their ideal temperature, which is 12–15°C (55–60°F); this is nowhere near as cold as a refrigerator and is more suited to the continuing ripening of cheeses. A cheesemonger will know not to wrap their cheeses tightly either, as this practice traps moisture and oxygen, promoting bacterial growth. Still-ripening whole cheeses should be kept unwrapped for air circulation, while mature ones should be wrapped in waxed paper. They will carefully tend their cheeses until they are at peak condition; if not, good cheese sellers will know to either not sell the cheese or to sell it with instructions on how to ripen the cheese at home, if necessary (although carefully controlled ripening processes can be difficult, if not impossible, to achieve at home).

Cheese is a constantly changing product, and many cheeses go through various ripening stages during which they are perfectly edible — often your own preferences will dictate exactly how ripe or aged you like your cheese. You should only buy as much cheese as you need at a time; that way you won't risk wasting any. Over-ripe cheese will exhibit strong ammoniac or putrid smells. Ripe cheeses can smell very pungent, but the smell should never be unpleasant — if it is, then the cheese is probably past its best. If you must store cheese at home, remember that temperature and humidity are the two main factors that affect the condition of cheese; cheese likes temperatures between 7°C and 15°C (45°and 60°F) and relative humidity of about 80 per cent. Domestic refrigerators are too cold and too dry for cheese storage; they run at about 3°C (38°F) and 30 per cent humidity. Cheeses dry out in temperatures that are too cold. Cool cellars and sheds can be ideal places for cheese, as long as they are not damp or subject to warming during the day. They should also be dark — cheese prefers to be kept in the dark. If the refrigerator is your only option, then keep cheese in the vegetable compartment, which is less cool and more humid than other areas.

Older cheeses keep for longer than younger cheeses. Wrap hard and semi-hard cheeses loosely in plastic wrap, on the exposed parts only, leaving the crusts free to breathe. Softer cheeses (such as surface-ripened ones) should be wrapped in waxed paper, then in plastic wrap, and not too tightly, as the cheese needs to breathe. Wrap blue cheeses in foil; these cheeses generally have no crust to protect them and are subject to moisture loss (they can lose up to 30 per cent of their weight by liquid oozing out or evaporating). Ideally, all cheese wrappings should be changed daily.

COOKING WITH CHEESE

Cheese is a versatile ingredient, but different cheeses react differently to heating, depending largely on their moisture content. Harder cheeses need a higher cooking temperature to melt than do soft or semi-soft cheeses. The longer cheese is cooked, the harder and drier it becomes, so it is important not to overcook dishes containing cheese. If using cheese in a sauce or soup, grate it finely, add it last thing, and do not boil the sauce or soup thereafter. Avoid over-stirring also, which might push all the cheese together to form a gooey lump. Acid, such as that found in lemon juice or white wine, will 'loosen' a clumpy, over-heated cheese mixture. Avoid over-cooking cheese on gratins or pizzas, otherwise it will become tough.

Cheeses that are curdled using acid rather than rennet will not melt when heated, but instead will keep their shape and simply dry out; ricotta and most fresh goat's cheeses fall into this category. Some cheeses become stringy when melted, as their protein fibres form long, rope-like threads that can both stretch and stick to each other; mozzarella, emmental and cheddar are examples. Cheeses that have had their protein fibres broken down through deliberate enzyme action, such as parmesan and pecorino, disintegrate into very small bits when heated, making them the perfect texture for pasta, risotto and polenta dishes as well as soups and sauces.

GLOSSARY OF CHEESE TERMS

Affine A French term for finishing or curing a cheese. The person who carries out the curing process is called an *affineur*.

Aged An aged cheese is one cured for longer than 6 months. Aged cheeses are fuller flavoured and sharper tasting than those that are not aged. Aging is done in special rooms where temperature and humidity are scrupulously controlled. Under these conditions, the micro-organisms that give various cheeses their unique characteristics flourish, helping flavour and texture to develop. Aging is also called curing.

Annatto This is a tasteless, odourless, vegetable-derived dye used to give certain cheeses a deep, orange-yellow paste or rind. It is not a preservative.

Ammoniac A term used to describe cheese that tastes and/or smells of ammonia — either a slight whiff or a very strong presence. The cause is usually over-ripeness or bad handling (a cheese that has been subject to fluctuating temperatures, for example). White-mould cheeses in particular taste unpleasantly ammoniac when over-ripe.

A.O.C. Stands for Appellation d'Origine Contrôlée, a French system whereby a particular cheese is guaranteed to come from a specific region and be made according to the specifications traditional to that cheese. In France, some 35 cheeses are protected by an A.O.C. designation and are marked as such on their packaging. Other European countries have a similar system, including the Italian Denominazione d'Origine Protetta, or D.O.C., and the Spanish D.O., which stands for Denominación de Origen. Such regulations specify things such as the animal's diet — for example, animals can only eat grasses and flowers, supplemented in winter by hay from local pastures.

Aroma The smell of cheese varies hugely depending upon the variety. Words used to commonly describe the aromas of cheese are 'fruity', 'nutty', 'earthy', 'mushroomy', 'sweet', 'clean', 'lemony', 'tangy' or 'barnyardy'. Sometimes the exterior of a cheese has completely different aromas to the interior.

Artisanal Artisanal cheeses are those made in small quantities, often using milk from just a few farms. As a result, such cheeses usually display unique characteristics, for which they are sought out. Such cheeses can generally be relied on to be carefully made and of high quality.

Bloomy rind The white, soft edible rind produced when white penicillium mould grows on the surface of certain cheeses (such as camembert, brie and some chèvres). This mould is sprayed over the surface before curing.

Blue (or blue vein) A descriptive term applied to the network of blue-green mould streaks allowed to flourish throughout the interior of blue cheeses, the most famous examples being roquefort, gorgonzola and stilton. Such cheeses are commonly (though not invariably) strong in both aroma and flavour, except for blue-brie hybrids, which are much milder.

Brine A salt-and-water solution in which certain cheeses are washed or dipped; some cheeses (such as feta) are stored in brine.

Casein The principal protein in milk. During cheese making, casein solidifies or curdles through the addition of a coagulant (such as rennet).

Culture A starter (usually a percentage of lactic acid, bacteria, mould spores, enzymes and other micro-organisms) which determines the curdling process and converts lactose into lactic acid. Cultures also give specific flavours and characteristics to a cheese variety.

Curdling An early part of the cheese making process, when the milk proteins are formed into curds by the action of rennet or acids.

Curing Also called aging or ripening, curing is the method by which a cheese acquires its distinct qualities, being a process carried out under very controlled conditions of temperature and humidity.

Degree of hardness The easiest way to categorise cheeses is by how hard they are. This determination is to do with the amount of moisture and fat in each one — moisture and fat control texture and firmness.

Double cream Describes cheeses with at least 60 per cent butterfat content.

Farm cheese Cheese made on a farm using milk from that farm only.

Grana An Italian term used to describe the hard, granular texture of certain hard cheeses, notably parmesan, romana, sapsago and asiago.

Hard cheese A classification applied to those cheeses that have a moisture content of about 34 per cent and a butterfat content of about 50 per cent. Examples are parmesan and aged pecorino.

Lactic acid bacteria Organisms used in starter cultures in cheese making.

Lactose The sugar naturally present in milk.

Medium-aged Describes semi-firm, firm or hard cheeses that have been cured for 3–6 months; their flavours are mellow and smooth.

Mould Fungi that are deliberately grown on (in the case of such cheeses as brie or camembert) or in (as with blue cheeses) a cheese during curing. The various moulds give the cheese certain desirable characteristics of taste, texture and aroma, and should not be unpleasant or overly ammoniac. The hoop or container in which a cheese is formed is also called a mould.

Mouth feel To do with texture, mouth feel is literally about how a cheese feels in your mouth, whether creamy, crumbly, grainy, dense, smooth etc.

Natural rind An exterior rind that forms naturally as the cheese ages, but without the aid of ripening aids, such as washing. Most semi-hard and hard cheeses have natural rinds (emmental, cheddar and parmesan, for example).

Over-ripe Describing a cheese that is past its prime and not good to eat; flavours and textures are over-developed and 'off' and the bacteria have died.

Pasta filata An Italian term meaning to 'spin paste' or 'spin threads'. Some Italian cheeses are made by heating and then stretching the curds and finally moulding them into a particular shape. Such cheeses are characteristically stringy and elastic when heated; two classic examples are mozzarella and provolone. Such cheeses are also called 'plastic curd' cheeses.

Paste The name given to the interior of soft-ripened cheeses, such as brie and camembert.

Pasteurisation The process whereby milk is heated to a certain temperature in order to kill disease-causing pathogens. Unfortunately, 'good' bacteria that are useful for starting fermentation are also killed.

Processed cheese This is made from a combination of cheese and other ingredients such as enzymes and stablisers. Processed cheese has an extended shelf life and typically a very uniform, 'plastic' texture and mild, sweet taste. Avoid such cheeses altogether, as they are a wholly industrial product and nothing like 'real' cheese in any way, shape or form.

Rennet An enzyme used to coagulate (or 'curdle') milk during the initial stages of cheese making. It causes the milk to separate into curds (clumpy solids) and whey (liquid). Rennet also contributes to the flavour of cheese, aiding in the release of peptides during maturation. It is traditionally sourced from the fourth stomach of new-born calves, although large-scale cheese manufacturers use rennet made from yeast or bacteria.

Rind The external surface of a cheese, which can be thick, thin, waxed, greased, hard, soft, mouldy or sticky, depending on the type of cheese. It both protects the interior and contributes to the flavour.

Ripe Describing a cheese that is at the end of its aging process and in peak condition for eating. Ripeness varies widely between types of cheeses.

Semi-hard A category of cheeses based on texture and moisture content; examples are cheddar, colby, edam and gouda.

Semi-soft Another category of cheeses based on texture and moisture content — a wide variety of cheeses fall into this category, including fontina, monterey jack, havarti and muenster. These cheeses all tend to melt well, so they are suitable to cook with.

Sharp Describing mature cheeses (mainly applied to semi-hard, hard and blue-veined varieties) that have a well-developed, biting flavour.

Smoked cheese Some cheeses (such as cheddar, provolone, mozzarella) are smoked to give a singular flavour. The traditional method involves slow smoking over woodchips, although these days it is more likely that liquid smoke (a flavouring made by collecting wood smoke) will have been added to the brine.

Soft or fresh cheeses Very young cheeses with a high moisture content, such as cottage and cream cheese, mascarpone and ricotta. They are made using lactic acid cultures, giving them a pleasant tangy flavour, and are meant to be eaten within a few days or weeks of making.

Solids All the components of a cheese other than water, including the proteins, milk fat and milk sugars.

Stabilisers Substances (such as gelatine, xanthan gum, guar gum and locust bean gum) added to cheeses to improve their consistency by binding water to solid matter. Whey protein concentrate is also used as a stabiliser.

Starter Lactic acid, bacteria or mould spores, enzymes or other natural cultures that are added to milk to curdle it by converting lactose (milk sugars) into lactic acid, thus starting the cheese-making process.

Surface-ripened Refers to cheese that ripens from the exterior first, then through to the interior; ripening is caused by mould, yeast or bacteria that are deliberately applied to the surface. When fully ripe, such cheeses have soft, creamy, sometimes oozing interiors. Mould-ripened cheeses include brie and camembert; two washed-rind cheeses are taleggio and pont l'évêque.

Washed rind Describes a type of cheese whose rind has been regularly washed with brine, beer, cider, wine, brandy or even oil while maturing, to keep the cheese moist and encourage the growth of flavour-forming bacteria, which help ripen the cheese. Washed-rind cheeses are a type of 'surface-ripened' cheese; that is, they ripen from the outside in.

Whey The watery part of milk that separates out from the curds when milk is first curdled with rennet or lactic acid. It contains the majority of the lactose, or milk sugar.

ASSEMBLING A CHEESE BOARD

A cheese board should be the easiest thing to assemble, but many people, whether through lack of confidence or simply not understanding a few basic pointers, either get it completely wrong or avoid the cheese course altogether. This is a great pity, really, as there are few things more appreciated by guests than sharing a carefully selected array of cheeses, or a single cheese at peak ripeness, with a few suitable accompaniments.

If you are uncertain where to begin, allow yourself to be guided by a good cheese merchant or assistant at a deli cheese counter — tell them what you like, explain what will precede the cheese course and follow their advice according to what is in stock and in best condition. It's important to decide whether to serve your cheese course before dessert, as is the French custom, or after it, as do the British. If before, then you can serve wines from dinner with your initial cheese selection(s) then move on to a sweeter, dessert wine that is equally suited to stronger cheeses, such as blue cheeses. Soft, mild cheeses are best eaten first and stronger ones last.

You can theme a selection — for example, if you are serving Italian or Spanish food, consider serving cheese from those countries, or those similar in style. Alternatively, you could choose to serve all goat's or sheep's milk cheeses, choosing from a variety of textures, ages and flavours from within those cheese families. Another option is to offer a classic selection that runs the gamut of soft, surface-ripened, hard or semi-hard and blue. It is best to not serve too many types (three to four is sufficient). Or simply make a feature of just one cheese, such as a wedge of crumbly, aged cheddar, a whole brie or washed rind, or a generous chunk of something blue.

Under no circumstances should you pre-cut or slice the cheeses into small pieces; serve them in the pieces in which they were purchased (unless they are enormous, of course, such as a very large piece of parmesan, and you only want to offer a portion.) Allow 100–150 g (3½–5 oz) of cheese per guest, perhaps a little more if dinner was light and there is no dessert.

Have your cheeses at room temperature for serving to optimise their flavour; take them out of the fridge at least an hour before serving. If cutting a chunk from a larger piece, do this at the last minute so it doesn't dry out (or, in the case of a whole surface-ripened cheese, so it doesn't ooze out). Place the pieces of cheese on the serving plate with the rind side outermost.

Small knives and plates should be supplied to each person sharing the plate, and accompaniments (see below) passed separately.

CHEESE ACCOMPANIMENTS

The flavours and textures of cheese are complemented by a whole range of accompaniments, depending on the type of cheese and when and how you are serving it. Fruits such as pears, apples, figs and grapes, with their mellow, sweet-acid balance and juicy textures, are all divine with cheese; do not serve fruits such as strawberries or other berries, pineapple, pawpaw, melon, citrus or bananas with cheese, as the flavours and textures of these will not complement it. (The exception to this is watermelon with feta, which is a classic Greek partnership.)

Plums, apricots and peaches also work with certain cheeses, such as surface-ripened types, ricotta and other fresh cheeses and mild blues. Dried fruits — figs and muscatels in particular — are excellent with most cheeses (although not, perhaps, the majority of goat's cheeses), as are nuts. Lightly toasted almonds, walnuts and hazelnuts are all good choices, but make sure these are the freshest possible — it is nice to offer them in their shells, if appropriate, and crack them at the table, where they will be appreciated with cheddars and other mature, semi-hard and hard cheeses.

In Europe, certain creamy cheeses, such as cabécou and gorgonzola, are served drizzled with a mixture of honey and walnuts.

Look in your deli or cheese shop for fruit pastes or logs, many of which are designed to go with cheese and are made from fresh or dried fruits. Quince, plum or apple pastes and fig and walnut logs are among the more common and most popular. Cooked fruit compotes (pears or prunes cooked in port or red wine, for example, or poached quinces) or roasted pears, figs, apricots or apples are also wonderful companions to cheeses. In the south of France, it is not unusual to be served a sweet cherry compote with the local goat's cheese. Fresh dates, with their intense, caramelly sweetness and rich texture, go surprisingly well with creamy cheeses such as blue brie, triple cream cheeses, and other surface-ripened specimens.

Bread, of course, goes superbly with cheese, and fresh baguette is always an appropriate choice. Heavy nut bread (walnut, for example) is excellent with goat's cheese, triple cream, blue cheese or cheddar, and dense dried fruit breads (such as raisin bread) go well with many cheeses, including washed rind varieties, brie and camembert, blue cheese, cheddar, ricotta or fromage blanc. Try heavy rye bread with aged gruyère, gouda, edam or other mature semi-hard cheeses.

Olives taste best with fresh cheeses, goat's cheese and some sheep's milk cheeses. Sometimes, especially if serving your cheese for lunch, a light, crunchy green salad strewn with a few nuts and dressed with a nut oil dressing is a fabulous accompaniment to a wedge of cheese.

Crackers can be excellent too, but beware of many commercial varieties, which are too heavily flavoured and salty to be successfully served with cheese. Oatcakes are a marvellous complement to cheddar, aged gruyère and blue cheeses too.

milk, cream and butter

Milk is one of the oldest of all foods; sheep and goats were first domesticated about 11,000 years ago in the area that is now modern-day Iran and Afghanistan, and by about 7000 BC cows were being herded in Turkey and parts of Africa. It is thought that milk very quickly became an important food for humans, providing vital calcium and protein.

The Greeks and Romans knew all about using milk for cheese and there is archeological evidence that milk was used in Neolithic-era England. While we tend to think of 'milk' as cow's milk, different types of milk are common in various parts of the world — reindeer, camel, water buffalo and yak, as well as the more familiar goat and sheep milks, are among these other types. Cow's milk, though, is held to account for about 90 per cent of the milk consumed worldwide.

Milk is a pale, opaque liquid produced in the mammary glands of female mammals — in fact, the ability to produce milk is a defining characteristic of all mammals. It is used to nourish newborn offspring and is composed of water, protein, calcium, vitamins (notably B2 and small quantities of C) and saturated fat. Milk is not just a useful foodstuff in its own right but is also the base ingredient for many other foods. From yoghurt, butter and cheese to cream and milk powder, this simple food has spawned the entire category that we call 'dairy' food. With its associated bucolic images of cows grazing on green, lush pasture, milk has come to symbolise purity, wholesomeness and goodness, and is often marketed as 'nature's perfect food'. Some might find this, these days at least, an unsettling claim, given that the majority of milk we purchase in the Western world has undergone a great deal of processing. Once, milk was a more varied product, its flavour governed by the breed of cow it came from, the diet of that cow and the season in which it was produced. Most of our milk nowadays comes from a single breed of cow, the black and white Holstein-Friesian, favoured by the dairy industry for its ability to produce prodigious amounts of milk every day.

Milk is largely homogenised before sale, by being forced, at very high pressure, through minuscule pipes. This disperses the fat globules evenly throughout the milk and stops them from rising to the top (and forming cream). Homogenised milk is whiter in colour and milder in flavour than unhomogenised milk and has a thinner mouth feel. Like many foodstuffs, milk has become an industrial product, with little of the character or natural variations that once were typical. Unless one has access to raw milk from a farm, we are unlikely to encounter milk in its original, unadulterated form and in the main, the rich taste of untreated milk is a thing of the past.

HOW MILK IS PRODUCED

There is not much about milk, that most 'everyday' of substances, that is really terribly everyday at all — especially not the way in which it is miraculously produced in the body of an animal. Cow's milk, for example, begins its life as green pasture, which an average cow spends eight hours a day eating. It takes about 8 kg (17½ lb) of grass to make just 1 litre (35 fl oz/4 cups) of milk. The grass undergoes a complex process of mastication, rumination (the re-chewing of regurgitated grass), and a journey through the cow's four stomach chambers, before it is totally digested.

Milk is made in the cow's udder, which comprises four compartments separated by strong membranes. Blood passes though the udder and components in it are turned into milk, which is secreted through the teats. A network of blood and lymph vessels in the udder feeds blood, via capillaries, into milk-producing cells; about 90,000 litres (24,000 gallons) of blood pass through a cow's udder every day and it takes some 800 litres (210 gallons) of blood to produce 1 litre (4 cups) of milk. A full cow's udder can hold up to 40 litres (10½ gallons) of milk.

Milk, whether from a cow, goat or sheep, has a fairly consistent nutritional profile, although percentages of nutrients vary between species. Cow's milk contains about 4 per cent fat (although this can vary between breeds; the Holstein-Friesian produces milk with about 3.6 per cent fat, while the Jersey has milk of about 5.2 per cent fat), as does goat's milk. Sheep's milk has a whopping 7.5 per cent fat content. Milk also contains proteins, various minerals and lactose, the carbohydrate component of milk. Lactose accounts for about 40 per cent of the calories in cow's milk and also for its sweet taste. Humans require a special enzyme to digest lactose, which some people lack; these people are 'lactose intolerant'.

There are special lactic-acid bacteria, notably lactobacilli and lactococci, that thrive on lactose and convert it into lactic acid; this acid makes the milk less habitable for other bacteria. It is these bacteria which, throughout the eons before factory-controlled processing, soured milk but still kept it drinkable, and by souring it made it ready to be transformed in cultured butter, cheeses, yoghurt and the like.

PASTEURISATION

Harmful organisms (including bacteria, yeasts and moulds) can easily flourish in milk, especially if it has not come from a sanitary milking environment or has been handled or stored unhygenically. Since the early to mid-twentieth century, milk has been routinely pasteurised. This heating process, pioneered in the nineteenth century by the French scientist Louis Pasteur, was not designed to kill all the organisms in milk — this would be called 'sterilisation' — but rather to kill off the bacteria that cause spoilage. It does, however, eliminate all but about 1 per cent of the bacteria in milk; along with the 'bad bugs', it also wipes out immunoglobins, enzymes such as lipase and phosphatase (which give flavour and aroma to cheeses), and some 20 per cent of milk's vitamin B6.

Pasteurisation increases the refrigerated shelf-life of milk to two to three weeks. The process is simple: milk is heated to 63°C (145.5°F) and held at that temperature for a period, and then rapidly cooled to 4°C (39°F). In another method, called Ultra Heat Treating, the milk is brought to a very high temperature (138°C/280°F) for a fraction of a second and then rapidly cooled; this milk will keep, refrigerated, for some 2–3 months. It will even keep, unrefrigerated and unopened, for very much longer than this.

There are some who disagree with the routine pasteurisation of milk, citing the damage it does to the beneficial, living organisms in milk (the process destroys *Lactobacillus acidophilus*, for example, which is used to make yoghurt and is a 'good' bacterium, believed to aid digestion). Critics and cheese experts say that using pasteurised milk gives cheese a 'cooked' flavour and contributes to rubbery textures. Advocates of raw-milk cheeses say that you simply can't make cheese from 'off', bacteria-infected milk — it just won't look, develop or taste right — and that clean, raw milk is a very safe substance. Others believe that drinking raw milk is dangerous; indeed, people do become sick as a result of consuming it, although the chances of this happening are negligible, provided the milk was sourced and stored in a hygienic manner. In the United States, some 28 states allow the sale of raw milk, and in the United Kingdom, about 200 suppliers are allowed by law to supply raw milk.

Whichever side of the raw-milk fence one camps on, however, it is a fact that the dictates of modern, mass production, where milk is sourced from hundreds or thousands of farms and stored in common vats before use, makes pasteurisation a necessity. Just one small quantity of contaminated milk will spoil whatever other milk it is mixed with, and milk, butter and cheese factories cannot take that risk.

MODERN MILK PRODUCTION

Over the past 30 or 40 years, dairy farming has changed. Once carried out on small farms, it is now dominated by extremely large-scale, mechanised operations — there are conglomerate dairies in the United States where cows in their thousands are milked. In that country, many dairy cows are kept under cover, rather than on pasture. Size of operation is not the only thing that has changed in recent times; cows have been specially bred to produce up to three times as much milk as they did 50 years ago. Many cows are given a growth hormone called BST (banned in Canada and the European Union) to further stimulate milk production, and giving cows these frequent injections does harm to their health (for example, they become more prone to mastitis).

Cows only produce milk when they have given birth so, to keep them in constant milk production, cows are artificially inseminated so they will give birth every year. The calves are quickly taken from the cow, so her milk can be used for production rather than go to the calf; animal advocates argue that such separation is a cruel practice. Modern dairy cattle are managed in such a way to provide an unnaturally constant, and large, amount of milk, and some people argue that the stress this places on the animals is extreme.

Humans tamper with milk beyond homogenising and pasteurising it, too. Milk companies offer milks with various levels of fat content, from full fat (about 3.5 per cent) down to practically zero fat. The fat is removed by centrifuging the milk, which renders the milk fairly thin and uninteresting; the recipes in this book require full-fat milk. Milk with some or all of the fat removed behaves differently when cooked, lacking the correct structure and body (not to mention flavour) to be successfully used in recipes requiring whole milk. Low-fat milks may also have dried milk proteins added to them to give them more body, and this can make them taste oddly stale.

CREAM

Milk is an emulsion of fats suspended in liquid. When left to stand, most of the fat in the milk, which is less dense than the water in the milk, rises to the top. This thick, rich substance is cream. The cream takes up to 24 hours to separate naturally from the milk, although since the late nineteenth century this process has been carried out by centrifuge. The process of removing the layer of cream from the top of the milk is called 'skimming', hence the term 'skimmed milk' to describe milk with the cream removed.

Cream has about 30 per cent fat. Milk contains an almost even ratio (by weight) of fats and proteins, whereas in cream, the fats outweigh protein by about ten to one; this is why cream does not 'split' or 'separate' when heated, as it is the proteins that cause this to happen.

Cream is the basis of butter and is also used in cheese making. It can be boiled without curdling, and boiling evaporates the moisture, thickening the cream, a process that is often employed in sauce making. Cream can be whipped, or processed into many other cream products, sour cream, crème fraîche and clotted cream among them. Clotted cream, an English speciality, is made by 'cooking' unpasteurised cream in shallow metal pans over low heat for about 12 hours, during which time it acquires a slightly caramelised taste, a clotted texture and a thin yellow crust. It has a fat content of about 60 per cent and is traditionally served with scones and jam as a tea-time treat. Clotted cream cannot be made from pasteurised cream. Like milk, however, the cream we routinely buy in supermarkets has been pasteurised, although not generally homogenised, as this makes it very hard to whip.

TYPES OF CREAM

The regular use of cream as an ingredient is only as old as refrigeration, as cream spoils rapidly and was traditionally made into butter to preserve it. 'Double' cream has a 48 per cent fat content and is very thick. 'Single' cream is not pure cream at all, having milk added to it to make it 10–12 per cent fat. Use pure cream (called 'heavy' cream in America) for all your cooking and whipping needs unless a recipe specifies otherwise. 'Thickened' cream should be avoided as it has been artificially thickened with vegetable gums. One cup of cream will make 2 cups of whipped cream; cream must have a fat content of 30 per cent in order to whip. Use cold cream for whipping, as room temperature cream will not whip well. Use a deep bowl; this will shorten the whisking time. Cream can be whipped by hand, using a balloon whisk or egg beaters, or with electric beaters. If using electric beaters, take great care, as cream can very easily be over-whipped and become unusable for anything except homemade butter.

EVAPORATED AND CONDENSED MILK

The idea of 'cooking' milk to concentrate and thus preserve it is an ancient one, and is still very prevalent in India. Simply by boiling, milk can be cooked down by evaporation into a thick brownish paste containing about 10 per cent moisture and 20 per cent fat.

Tinned milk dates back to 1852. An American dairy farmer, Gail Bordern, was travelling by ship from Britain to the United States. Extremely rough seas made the cows, kept on board to supply the ship with fresh milk, sea sick and they were unable to produce. As a result of the lack of milk, infants on board became ill. In those pre-refrigeration days, milk spoiled extremely quickly as there were no safe, sanitary ways to store it. Upon his return home, Bordern began to experiment with ways to make milk storable, and discovered that milk was made of 87 per cent water. He discovered that if he boiled the milk to evaporate off some of the water, then stored the milk under airtight conditions, it would resist deterioration. By adding sugar to the milk, and removing all the milk's fat, he found he could make it even more bacteria proof and, in 1864, Bordern started producing Eagle Brand condensed milk.

BUTTER

Butter as a food is probably as old as milk production itself. Historians claim that the Scythians, several hundred years before the birth of Christ, valued butter so much that they employed blind slaves to make it so they wouldn't be distracted from their important task. Butter was around for thousands of years before the Scythians, though; it is mentioned in ancient Egyptian texts. Later, the Romans used butter to dress battle wounds, and as a beauty treatment for their skin and hair.

In all probability, the butter made by the ancient Mediterranean and Near Eastern cultures was ghee, not butter as we know it, as ghee keeps well in hot climates; pure butter is very perishable. It was also no doubt made from goat's or water buffalo's milk, rather than the now more familiar cow's milk. In cooler, northern climes, butter was made in the solid form we know today; the Scandinavians were making butter for export in at least the twelfth century. Throughout Europe, from the fall of the Roman Empire to throughout the Middle Ages, butter was a very common food, eaten mainly, it would seem, by the lower classes and not held in particularly high regard. This changed when, in the sixteenth century, the Catholic Church allowed butter to be consumed during the pre-Easter Lent period, and bread and butter then became *de rigueur* among the middle classes. Ireland and France became important centres of butter production. In France, butter became so popular that the country couldn't produce enough to satisfy demand, leading Emperor Napoleon III to offer a substantial prize for anyone who could formulate a butter substitute. A French chemist claimed the prize when, in 1869, he invented margarine, based on rendered beef fat and milk (vegetable-fat-based margarines came into production in the early twentieth century). The advent of margarine had an enormous impact on the consumption of butter, mainly because it was cheaper for the consumer to buy and also because, in the second half of the twentieth century, when animal fats in the diet came to be regarded as bad, it was seen as a healthier choice. This belief is changing, however, with new knowledge about the dangers of the trans fats and various additives contained in some margarines and a trend back to more 'natural' foods.

MAKING BUTTER

It is believed that ancient nomads were the first to make butter, and that they did so by mistake. Anyone who has ever over-whipped cream, to the point where clumps of fat form and liquid (buttermilk) separates out, will know that this is easy enough to do. Butter is simple to make. Even today, although mostly produced in huge volumes in high-tech factories, butter is made in much the same way it has been for millennia. Cream is churned until the fats separate from the liquid, the butter 'grains' are washed, salt is added as a preservative (although unsalted butter is also popular), air is removed, an amount of water is added, and the butter is shaped, packaged and sent to market.

Until the mid-nineteenth century, butter was made on farms, by hand. It was only made when sufficient cream had been accumulated; this might take a week or so, and in that time the waiting cream would turn sour. Cream fermented under such uncontrolled conditions could easily spoil, and often did. Today butter is still made from fermented cream, although the process of souring has long been made totally safe and hygienic. Butter is also commonly made with fresh, pasteurised cream. With the advent of centrifugal separators and dairy factories in the late 1800s, butter could be made more quickly and in vastly larger quantities. By 1900, more than half the entire butter output in the United States was produced in factories, with Europe being not far behind.

Butter comprises different percentages of fat (falling somewhere between 80 and 85 per cent), moisture and salt (up to 3 per cent), depending upon its type and where it is made. Many commercial manufacturers remove some of the fat so it conforms to the minimum amount required (in America, butter must have 80 per cent fat). It can also vary in colour from bright yellow through to nearly white, depending upon variables such as the time of the year it is made and the richness of the pasture. These days many manufacturers add colourants such as carotene or annatto to make it yellow. Butter remains solid when chilled, softens at room temperature, becomes spreadable at about 15°C (60°F) and has a melting point of 32–35°C (90–95°F). Butter keeps well, but should be refrigerated, as light and heat cause it to turn rancid within a few days. It absorbs odours, too, so keep it away from strong-smelling foods in the refrigerator, lest it become tainted. Butter freezes well; wrap it in a few layers of plastic wrap or place it in a zip-lock bag or other airtight container for up to 6 months. Don't wrap butter in kitchen foil, as the metal can encourage oxidisation in butter. When the exposed surfaces of butter turn a deep, translucent yellow, this is a sign that the butter is rancid — scrape such patches off and discard them.

TYPES OF BUTTER

Sweet cream butter The development of refrigeration in the nineteenth century and the invention of mechanised cream separation made it possible to make butter from fresh cream rather than fermented cream. These days, the vast majority of the butter made in the United Kingdom, United States, New Zealand and Australia is sweet cream butter. It has a higher moisture content than cultured butter, a lower burning temperature and contains more lactose. Sweet cream butter can be purchased either salted or unsalted — salt is added for flavour and as a preservative. Unsalted butter is generally preferred by bakers because of its more neutral flavour. Salted butter makes a good, all-purpose table butter, to use for spreading and low-temperature cooking.

Cultured or European-style butter This butter, made using soured cream, has a deep, complex, tangy flavour. Typically the fresh cream is soured by the addition of a culture, then left to stand at a controlled temperature for about 18 hours, during which time the characteristic, slightly acidic flavours develop as the live culture acts upon the cream. The cream is then churned into butter. Cultured butter has a lower moisture content than sweet cream butter, and a higher butterfat content. For these reasons it is preferred by professional pastry cooks (excess moisture in baking can adversely affect the results). It is more expensive to buy than sweet cream butter as the manufacturing process is costly; it is cheaper to make butter from fresh cream, as there is no waiting time involved. Cultured butter is sold salted or unsalted; unsalted is preferred for baking.

Ghee and clarified butter Ghee is the Indian word for clarified butter. To clarify butter, solid butter is melted and the pure butterfat is poured off, leaving the white milk solids and water behind, giving a product that is nearly 100 per cent milk fat. Without the milk solids, which easily burn, ghee and clarified butter have a much higher smoking point than ordinary butter, making them more suited for use as a frying medium. The main difference between ghee and clarified butter is that ghee is simmered for a while during the clarification process, to evaporate the moisture; during this time, the milk solids brown and impart a slightly nutty flavour to the finished product. Because they contain no moisture, ghee and clarified butter will keep much longer than regular butter.

To make clarified butter, place butter in a small, heavy-based pan and allow to melt slowly over low heat. Don't be tempted to turn up the heat to speed the process, or you may burn the solids. Once the butter has melted, leave it on the heat until a fine white scum forms on top; some of the solids will fall to the bottom but some will rise to the top. Carefully skim these off, then very carefully pour the yellow butter fat into a container, leaving behind the white liquid at the bottom of the saucepan. Store the clarified butter, covered to protect it from strong odours, in the refrigerator, where it will keep for at least 1 month. It will also keep, frozen, for 6–8 months. Ghee can be purchased in tins from Indian grocers and some supermarkets.

COOKING WITH BUTTER

Butter has myriad uses in the kitchen. From the simplest (using it to grease baking dishes) to the most complex (emulsifying it into fragile, elegant sauces such as hollandaise or béarnaise), it imparts wonderful flavours and textures that are not achieved with any other ingredient.

Butter picks up flavours well and is often mixed with other ingredients (such as herbs, garlic, strong cheeses, chopped sun-dried tomatoes, olives, saffron and other spices, lemon juice, wine reductions or capers) to form savoury spreads or 'compound' butters that are used in slices or as a pat over cooked meats, fish or vegetables.

Melted butter is often the basis of simple sauces used to enrich vegetable or pasta dishes, and sometimes these are made more complex by long cooking to deliberately caramelise the solids in the butter. This process produces the famed French *beurre noisette*, or 'nut-brown' butter, which, with its sweet, deep, nutty aromas, brings new depths of flavour to fish, egg and meat dishes. Care should be taken when cooking butter thus, as it can easily burn.

Butter is not suited to high-temperature cooking processes such as stir-frying and flash-frying, as it will burn and taste acrid. Butter is, however, indispensable in baking; it provides richness and texture and helps to provide the supporting structure in goods such as cakes and biscuits.

MARMITE DIEPPOISE
SERVES 6

450 ml (16 fl oz) cider or dry white wine

16 mussels, scrubbed and beards removed

50 g (1³/₄ oz) butter

1 garlic clove, crushed

2 spring onions (scallions), finely chopped

2 celery stalks, finely chopped

1 large leek, white part only, thinly sliced

250 g (9 oz/2³/₄ cups) sliced small button mushrooms

2 small bay leaves

300 g (10¹/₂ oz) salmon fillet, skinned and cut into chunks

400 g (14 oz) sole fillet, skinned and cut into chunks

12 large raw prawns (shrimp), peeled and deveined

300 ml (10¹/₂ fl oz) thick (double/heavy) cream

Pour the cider or white wine into a large saucepan and bring to a simmer. Add the mussels, cover the pan and cook for 3–5 minutes, shaking the pan occasionally. Place a fine sieve over a bowl then strain the mussels, reserving the liquid and discarding any mussels that haven't opened. Strain the liquid again, taking care to leave behind any grit or sand.

Clean the saucepan, add the butter and melt over a medium heat. Add the garlic, spring onion, celery and leek and cook for 7–10 minutes, or until the vegetables are just soft. Add the mushrooms and cook for a further 4–5 minutes, or until softened. While the vegetables are cooking, remove the mussels from their shells, discarding the shells.

Add the mussel liquid to the vegetables in the saucepan, add the bay leaves and bring to a simmer. Add the salmon, sole and prawns and cook for 3–4 minutes, or until the fish is opaque and the prawns have turned pink. Stir in the cream and cooked mussels and simmer gently for 2 minutes. Season to taste and serve.

MILLEFEUILLE OF LEEKS AND POACHED EGGS
SERVES 4

1 sheet pre-rolled frozen butter puff
  pastry, partially thawed
1 egg, lightly beaten
3 leeks, white part only
30 g (1 oz) butter
4 eggs

BEURRE BLANC
20 g ($^3/_4$ oz) butter
2 French shallots, finely chopped
3 tablespoons white wine
185 ml (6 fl oz/$^3/_4$ cup) chicken stock
175 g (6 oz/$^3/_4$ cup) unsalted butter,
  chilled and cut into cubes
white pepper, to taste
freshly ground black pepper

Preheat the oven to 190°C (375°F/Gas 5). Cut the pastry sheet diagonally into quarters. Place the triangles on a damp baking tray, lightly brush with the beaten egg and bake for 15 minutes, or until puffed and golden brown. Cool slightly then, using a sharp, serrated knife, slice the triangles in half horizontally and use a spoon to remove any uncooked dough from the inside.

Cut the leeks in half widthways and then into thin shreds. Melt the butter in a frying pan over medium heat, add the leeks and cook, stirring, for 10 minutes, or until they are tender. Season with salt.

To poach the eggs, bring a frying pan of water to the boil, then reduce the heat to low. Crack an egg into a saucer then slide the egg into the simmering water. Poach for 3 minutes, then remove carefully with a slotted spoon and drain on paper towels. Poach all the eggs in the same way.

To make the beurre blanc, melt the butter in a saucepan, add the shallots and cook, stirring, over medium-low heat for 4–5 minutes, or until soft but not browned. Add the wine, bring to the boil and cook over medium heat for 5–6 minutes, or until the liquid is reduced by half. Add the stock and continue boiling until the liquid is reduced by one-third. Reduce the heat to low then, whisking continuously, add the chilled butter, one piece at a time, making sure each is incorporated before adding the next, until the sauce has thickened. Take care not to overheat the sauce or it will separate; you may need to move the pan from the heat occasionally to avoid overheating. Season the beurre blanc with salt and white pepper to taste.

Divide the pastry bases among four warmed serving plates, top each with a spoonful of warm leek mixture, then a poached egg and a little beurre blanc. Cover each with a pastry top, season with black pepper and serve with the remaining beurre blanc drizzled on the side.

Jansson's temptation
SERVES 4

15 anchovy fillets
80 ml (2½ fl oz/⅓ cup) milk
60 g (2¼ oz/¼ cup) butter
2 large onions, thinly sliced
5 potatoes, peeled, cut into 5 mm
(¼ inch) slices, then into fine
matchsticks
500 ml (17 fl oz/2 cups) thick (double/
heavy) cream

Preheat the oven to 200°C (400°F/Gas 6). Soak the anchovies in the milk for 5 minutes, then drain well and rinse.

Melt half the butter in a large frying pan, add the onion then cook, stirring occasionally, over medium heat for 5 minutes, or until golden and tender. Chop the remaining butter into small cubes and set aside.

Spread half the potato over the base of a shallow ovenproof dish, top with the anchovies and onion and finish with the remaining potato.

Pour half the cream over the potato and scatter the reserved butter over. Bake for 20 minutes, or until golden. Pour the remaining cream over the top and cook for another 40 minutes, or until the potato is tender when pierced with the point of a knife. Season to taste with salt and pepper then serve immediately.

INDIAN DAL
SERVES 4–6

310 g (11 oz/1¼ cups) red lentils
2 tablespoons ghee
1 onion, finely chopped
2 garlic cloves, crushed
1 teaspoon grated fresh ginger
1 teaspoon ground turmeric
1 teaspoon garam masala
naan bread, to serve

Put the lentils in a large bowl and cover with water. Remove any floating particles and drain the lentils well.

Heat the ghee in a large saucepan, add the onion and cook, stirring, over medium heat for 4–5 minutes, or until soft. Add the garlic, ginger and spices and cook, stirring, for 1 minute, or until fragrant.

Add the lentils and 500 ml (17 fl oz/2 cups) water to the saucepan and bring to the boil. Lower the heat then simmer, stirring occasionally, for 15 minutes, or until all the water has been absorbed and the lentils are very tender, taking care that the mixture does not stick to the base of the pan.

Allow the mixture to cool slightly, then transfer to a serving bowl and serve warm or at room temperature with naan bread.

GNOCCHI ROMANA
SERVES 4

750 ml (26 fl oz/4 cups) milk
a pinch of ground nutmeg
200 g (7 oz/1²/₃ cups) fine semolina
3 egg yolks
65 g (2¼ oz/²/₃ cup) grated parmesan
  cheese
30 g (1 oz) butter, melted
80 ml (2½ fl oz/¹/₃ cup) pouring
  (whipping) cream
75 g (2½ oz/½ cup) grated mozzarella

Line a deep swiss roll (jelly roll) tin with baking paper. In a saucepan, combine the milk, nutmeg, salt and freshly ground black pepper to taste. Bring to the boil, reduce the heat and gradually stir in the semolina. Cook, stirring frequently, for 5–10 minutes, or until the mixture is very thick.

Remove the pan from the heat and cool slightly. Beat the egg yolks and half the parmesan together in a small bowl. Stir into the semolina, then spread the mixture in the tin. Cool slightly then refrigerate for 1 hour, or until the mixture is firm.

Preheat the oven to 180°C (350°F/Gas 4). Using a floured 4 cm (1½ inch) biscuit (cookie) cutter, cut the semolina into rounds and arrange in a greased shallow overproof dish. Pour the butter and cream over, then sprinkle with the combined mozzarella and remaining parmesan. Bake for 20–25 minutes, or until the gnocchi are golden and bubbling and heated through. Season with freshly ground black pepper and serve.

Zucchini with lemon and caper butter
Serves 4

LEMON AND CAPER BUTTER
100 g (3½ oz) butter, softened
2 tablespoons capers, rinsed, squeezed dry
   and chopped
2 teaspoons finely grated lemon zest
1 tablespoon lemon juice

8 small zucchini (courgettes) (about
   750 g/1 lb 10 oz), trimmed

To make the lemon and caper butter, combine the butter with the capers, lemon zest and juice in a large bowl. Using a wooden spoon, beat until well combined, then season to taste with salt and pepper.

Thinly slice the zucchini lengthways. Steam over boiling water for 3–4 minutes, or until just tender, then toss with the lemon and caper butter in a bowl and serve immediately.

Mustard seed butter
Makes about 100 g (3½ oz)

1 tablespoon vegetable oil
1 tablespoon mustard seeds
1 garlic clove, crushed
½ onion, diced
½ teaspoon Hungarian paprika
½ teaspoon sea salt
2 tablespoons chopped flat-leaf
   (Italian) parsley
2 teaspoons seeded mustard
90 g (3¼ oz/⅓ cup) butter, softened

Heat the oil in a small heavy-based saucepan and add the mustard seeds. Heat for about 1 minute, or until they start to pop. Add the garlic and onion and stir for about 5 minutes, or until softened. Cool slightly, season to taste with salt and freshly ground black pepper then transfer to a bowl, combine with the remaining ingredients and mix well. Serve on top of cooked meat, such as steak or lamb, or hot vegetables.

NOTE Mustard seed butter can be frozen; spoon the butter onto a piece of plastic wrap and roll into a log shape. Seal the ends and freeze for up to 3 months. To serve, simply cut slices off the frozen log; there is no need to thaw the butter before use if using it on top of cooked foods.

LOBSTER WITH FOAMING CITRUS BUTTER

SERVES 8

1 kg (2 lb 4 oz) live flathead lobsters
(see Tip)
sourdough or country-style bread, to serve

FOAMING CITRUS BUTTER
50 g (1³⁄₄ oz) butter
1 large garlic clove, crushed
1 tablespoon finely grated orange zest
1 tablespoon blood orange juice or regular
orange juice
1 tablespoon lemon juice
1 tablespoon finely snipped chives

Freeze the lobsters for 1–2 hours to render them unconscious. Heat the oven grill (broiler) to medium.

Plunge the lobsters into a large saucepan of boiling water for 2 minutes, then drain well. Using a sharp knife or cleaver, cut each lobster in half lengthways then place in a large baking tray, cut side up. Grill for 5–6 minutes, or until just cooked through.

Meanwhile, make the foaming citrus butter. Melt the butter in a small saucepan and when sizzling, add the garlic. Cook over medium heat for 1 minute, stirring often, then add the orange zest, orange juice and lemon juice and bring to the boil. Add the chives and season with salt and pepper.

Divide the lobsters among warmed serving plates and serve immediately, with the foaming citrus butter in small bowls for dipping.

TIP Instead of flathead lobsters you could use live slipper lobsters or crayfish for this recipe. You may need to boil them for a slightly longer time to cook them, and you'll need to remove the heads before slicing them down the middle. Depending on their size, they may also need a little longer under the grill (broiler).

CREAM OF SWEET POTATO AND PEAR SOUP
SERVES 4

20 g (¾ oz) butter
1 small white onion, finely chopped
5 medium orange sweet potatoes (about
    750 g/1 lb 10 oz), peeled and cut into
    2 cm (¾ inch) pieces
2 firm, ripe pears (about 500 g/1 lb 2 oz),
    peeled, cored and cut into 2 cm
    (¾ inch) pieces
750 ml (26 fl oz/3 cups) vegetable or
    chicken stock
250 ml (9 fl oz/1 cup) evaporated milk
mint leaves, to garnish

Melt the butter in a saucepan over medium heat, add the onion and cook, stirring often, for 3–4 minutes, or until softened.

Add the sweet potato and pear and cook, stirring, for 1–2 minutes. Add the stock and bring the mixture to a boil, reduce the heat to medium-low and cook for 20 minutes, or until the potatoes and pears are soft.

Allow to cool slightly, then transfer the mixture to a food processor, in batches if necessary, and process until smooth. Return to the saucepan, stir in the evaporated milk and gently warm over medium-low heat; do not allow the mixture to boil.

Season to taste with sea salt and freshly ground black pepper, then divide among warmed bowls. Scatter mint leaves over and serve immediately.

PASTA AND SPINACH TIMBALES
SERVES 6

500 g (1 lb 2 oz) English spinach, steamed
  and cooled
30 g (1 oz) butter
1 tablespoon olive oil
1 onion, chopped
8 eggs, lightly beaten
250 ml (9 oz/1 cup) pouring (whipping)
  cream
100 g (3½ oz) spaghetti or taglioni,
  cooked
60 g (2¼ oz/½ cup) grated
  cheddar cheese
50 g (1¾ oz/½ cup) freshly grated
  parmesan cheese
green salad and crusty bread, to serve

Preheat the oven to 180°C (350°F/Gas 4). Brush six 250 ml (9 oz/1 cup) ramekins or dariole moulds with some oil and line the bases with baking paper.

Using your hands, squeeze the spinach firmly to get rid of as much liquid as possible, then chop. Heat the butter and oil in a frying pan, add the onion and cook, stirring often, over medium-low heat for 8–10 minutes, or until the onion is softened. Add the spinach and cook for 1 minute. Remove from the heat and transfer to a bowl to cool.

Whisk in the eggs and cream, stir in the spaghetti and grated cheeses then season to taste with salt and freshly ground black pepper. Stir to combine well and spoon into the prepared ramekins.

Place the ramekins in an ovenproof dish. Pour boiling water into the dish to come halfway up the sides of the ramekins then bake for 30–35 minutes, or until set and a small knife inserted into the centre of a timbale withdraws clean. Cover the ramekins with foil if the timbales brown too quickly.

Allow the timbales to rest for 15 minutes then turn them out, running the point of a small knife around the edge of each ramekin to loosen. Invert the timbales onto serving plate and serve immediately with green salad and crusty bread.

Spaghetti carbonara
SERVES 6

500 g (1 lb 2 oz) dried spaghetti
8 rindless bacon slices (about 450 g/1 lb)
2 teaspoons olive oil
4 eggs, lightly beaten
50 g (1³/₄ oz/¹/₂ cup) freshly grated
   parmesan cheese
300 ml (10¹/₂ fl oz/1¹/₄ cups) pouring
   (whipping) cream
finely snipped chives, to garnish
   (optional)

Cook the spaghetti in a large saucepan of boiling salted water until *al dente*, then drain and return to the pan.

Meanwhile, cut the bacon into thin strips. Heat the olive oil in a heavy-based frying pan, add the bacon and cook over medium heat, stirring often, for 5–6 minutes, or until crisp. Remove and drain on paper towels.

Working quickly, whisk the eggs, parmesan and cream in a bowl until combined. Add the bacon, then pour the mixture over the hot pasta in the pan. Toss to coat the pasta with the sauce and cook over a very low heat for 1 minute, stirring, or until slightly thickened — take care not to overheat the sauce, or the eggs will scramble.

Season with freshly ground black pepper then divide among warmed bowls.

BÉARNAISE SAUCE
SERVES 6

1 French shallot, chopped
2 tablespoons dry white wine
2 tablespoons tarragon vinegar
3 sprigs tarragon
3 egg yolks
200 g (7 oz) butter, melted
1 tablespoon chopped tarragon leaves
1 tablespoon lemon juice
white pepper, to taste

Combine the shallot, wine, vinegar and tarragon sprigs in a small saucepan. Boil the mixture until only 1 tablespoon of the liquid remains. Strain and set aside.

Whisk the egg yolks with $1\frac{1}{2}$ tablespoons water and add to the saucepan. Place the pan over very low heat or over a pan of simmering water and continue to whisk until the sauce is thick. Take care not to overheat, or the eggs will scramble.

Remove the sauce from the heat, continue to whisk and slowly add the butter in a thin steady stream. Pass through a fine strainer, then stir in the chopped tarragon. Add the lemon juice to taste, and season with salt and white pepper.

Serve over beef steak, steamed asparagus, braised leeks or other simple vegetable dishes.

HOLLANDAISE SAUCE
SERVES 4

3 egg yolks
3 teaspoons lemon juice
150 g ($5\frac{1}{2}$ oz) unsalted butter, cut into cubes

Put the egg yolks and lemon juice in a saucepan over very low heat. Whisking continuously, add the butter piece by piece until the sauce thickens. Do not overheat, or the eggs will scramble. Season with salt and pepper.

The sauce should be of pouring consistency — if it is too thick, add 1–2 tablespoons of hot water to thin it a little.

Serve over poached eggs, steamed asparagus, braised leeks or other simple vegetable dishes.

Béarnaise sauce

TAGLIATELLE WITH VEAL, WINE AND CREAM
SERVES 6

500 g (1 lb 2 oz) thin veal steaks, cut into thin strips

plain (all-purpose) flour, seasoned, for dusting

60 g (2¼ oz/¼ cup) butter

1 onion, sliced

125 ml (4 fl oz/½ cup) dry white wine

125 ml (4 fl oz/½ cup) beef stock or chicken stock

170 ml (5½ fl oz/⅔ cup) pouring (whipping) cream

1 tablespoon grated parmesan cheese, plus extra to serve (optional)

600 g (1 lb 5 oz) fresh tagliatelle

flat-leaf (Italian) parsley, to garnish

Dust the veal strips with the seasoned flour, shaking off any excess. Melt the butter in a frying pan, then add the veal and cook over medium-high heat, stirring often, for 2–3 minutes, or until golden. Remove with a slotted spoon and set aside.

Add the onion to the pan and cook, stirring often, for 6 minutes, or until soft and golden. Pour in the wine, bring to the boil and cook over high heat for 3–4 minutes, or until reduced by half. Add the stock and cream and season to taste. Bring the mixture back to the boil then cook over high heat for 3–4 minutes, or until reduced and thickened slightly. Add the veal and toss to coat, then stir in the parmesan.

Meanwhile, cook the tagliatelle in a large saucepan of boiling salted water for 3 minutes, or until *al dente*, then drain well. Divide the pasta among warmed bowls, pour the sauce over, scatter with parsley and serve immediately.

QUICHE LORRAINE
SERVES 8

TART PASTRY
220 g (7³/₄ oz/1³/₄ cups) plain
   (all-purpose) flour
150 g (5¹/₂ oz) unsalted butter, chilled and
   cut into cubes
1 egg yolk

FILLING
25 g (1 oz) butter
300 g (10¹/₂ oz) bacon, diced
250 ml (9 fl oz/1 cup) double (thick/
   heavy) cream
3 eggs
a pinch of grated nutmeg

Preheat the oven to 180°C (350°F/Gas 4).

To make the pastry, sift the flour and a pinch of salt into a large bowl, add the butter and rub in with your fingertips until the mixture resembles breadcrumbs. Add the egg yolk and a little cold water (2–3 teaspoons) and mix with a flat-bladed knife until the dough just starts to come together. Bring the dough together with your hands and shape into a ball. Wrap in plastic wrap and refrigerate for at least 30 minutes.

Roll out the pastry into a circle on a lightly floured surface and use it to line a 25 cm (10 inch) fluted loose-based tart tin. Trim, then pinch up the pastry edge to make an even border raised slightly above the rim of the tin. Slide onto a baking tray and rest in the refrigerator for 10 minutes.

Line the pastry shell with a piece of greaseproof paper and baking beads (use dried beans or rice if you don't have beads). Blind bake the pastry for 10 minutes, remove the paper and beads and bake for a further 3–5 minutes, or until the pastry is just cooked but still very pale.

To make the filling, melt the butter in a small frying pan and cook the bacon until golden. Drain on paper towels.

Mix together the cream and eggs and season with salt, pepper and the nutmeg. Scatter the bacon into the pastry shell and then pour in the egg mixture. Bake for 30 minutes, or until the filling is set. Leave in the tin for 5 minutes before serving.

Salmon steaks with herb sauce
Serves 4

HERB SAUCE
375 ml (13 fl oz/1 1/2 cups) fish stock
125 ml (4 fl oz/1/2 cup) white wine
3 tablespoons snipped chives
3 tablespoons chopped flat-leaf (Italian)
  parsley
2 tablespoons chopped basil
2 tablespoons chopped tarragon
250 ml (9 fl oz/1 cup) pouring (whipping)
  cream
2 egg yolks

2 tablespoons olive oil
4 salmon steaks, skinned (about 250 g/
  9 oz each)

To make the herb sauce, combine the stock and wine in a saucepan and bring to the boil. Boil for 5 minutes, or until the liquid has reduced by half. Transfer to a food processor, add the chives, parsley, basil and tarragon and process for 30 seconds. Return the mixture to the pan, then stir in the cream and bring to the boil. Reduce the heat to low and simmer for 5 minutes, or until reduced by half. Remove from the heat.

Place the egg yolks in a food processor and process until smooth. With the motor running, add the herb mixture in a thin, steady stream and process until smooth, then season to taste with salt and freshly ground black pepper.

Heat the oil in a frying pan, add the salmon steaks and season to taste. Cook over medium-high heat for 3 minutes each side, or until just cooked through; the salmon should still be a little pink in the middle.

Serve the salmon immediately with the herb sauce spooned over.

CHICKEN AND BACON GOUGÈRE
SERVES 6

60 g (2¼ oz/¼ cup) butter
1–2 garlic cloves, crushed
1 red onion, chopped
3 bacon slices, chopped
30 g (1 oz/¼ cup) plain (all-purpose) flour
375 ml (13 fl oz/1½ cups) milk
125 ml (4 fl oz/½ cup) pouring
  (whipping) cream
2 teaspoons wholegrain mustard
250 g (9 oz) cooked chicken, chopped
1 large handful parsley, chopped

CHOUX PASTRY
60 g (2¼ oz/½ cup) plain (all-purpose)
  flour
60 g (2¼ oz) butter, chilled and cut into
  cubes
2 eggs, lightly beaten
35 g (1¼ oz/⅓ cup) freshly grated
  parmesan cheese, plus extra to serve
  (optional)

Preheat the oven to 210°C (415°F/Gas 6–7). Grease a deep 23 cm (9 inch) ovenproof dish.

Melt the butter in a frying pan, add the garlic, onion and bacon and cook for 5–7 minutes, stirring occasionally, or until cooked but not browned. Stir in the flour and cook for 1 minute, then gradually add the milk and stir until the mixture boils and thickens. Reduce the heat and simmer, stirring, for 2 minutes, then add the cream and mustard. Remove from the heat and stir in the chicken and parsley. Season with pepper.

To make the choux pastry, sift the flour onto a piece of baking paper. Put the butter in a large saucepan with 125 ml (4 fl oz/½ cup) water and bring to the boil over medium heat. Remove from the heat and add the flour all at once, stirring vigorously with a wooden spoon. Return to the heat and continue beating until the mixture forms a ball that leaves the side of the pan. Transfer to a large clean bowl and cool slightly. Gradually add the beaten egg, about 3 teaspoons at a time, beating constantly until the mixture is thick and glossy; it should fall heavily from the spoon; you may not need all the egg. Stir in the parmesan.

Pour the filling into the prepared dish and spoon heaped tablespoons of choux around the outside. Bake for 10 minutes, then reduce the oven to 180°C (350°F/Gas 4). Bake for 20 minutes, or until the choux is puffed and golden. Sprinkle with a little more grated parmesan if desired, then serve immediately.

Berry trifle
SERVES 8–10

550 g (1 lb 4 oz/1³/₄ cups) redcurrant jelly
170 ml (5¹/₂ fl oz/²/₃ cup) orange juice
625 ml (21¹/₂ fl oz/2¹/₂ cups) pouring
    (whipping) cream
250 g (9 oz) mascarpone cheese
30 g (1 oz/¹/₄ cup) icing (confectioners')
    sugar
1 teaspoon natural vanilla extract
¹/₄ teaspoon ground cinnamon
250 g (9 oz) savoiardi (lady fingers/sponge
    finger biscuits)
375 ml (13 fl oz/1¹/₂ cups) marsala
200 g (7 oz) fresh raspberries
200 g (7 oz) fresh redcurrants
200 g (7 oz) large fresh strawberries,
    hulled and quartered
200 g (7 oz) fresh blueberries
200 g (7 oz) fresh blackberries (see Note)

Melt the redcurrant jelly in a small saucepan over medium heat. Remove from the heat, stir in the orange juice and set aside until the mixture cools to room temperature.

Using electric beaters, beat the cream, mascarpone, icing sugar, vanilla and cinnamon in a bowl until soft peaks form.

Cut each biscuit in half widthways and dip each piece in the marsala. Arrange half the biscuits over the base of a 3.25 litre (110 fl oz/13 cup) serving bowl. Sprinkle a third of the combined berries over the biscuits and drizzle with half the remaining marsala and a third of the redcurrant sauce. Spoon half the cream mixture over the sauce. Repeat the layering with the remaining half of the dipped biscuits and marsala,
a third of the berries and sauce, and the remaining cream.

Arrange the remaining berries over the cream. Reserve the final third of the redcurrant sauce, cover and refrigerate. Cover the trifle with plastic wrap and refrigerate overnight. Before serving, pour the reserved redcurrant sauce over the berries to glaze. (Warm the sauce slightly if it has thickened too much.)

NOTE Any selection of fresh berries, to a total weight of 1 kg (2 lb 4 oz), can be used. Frozen berries are unsuitable for this recipe.

Eton mess
SERVES 4

4–6 ready-made meringues, about 15 cm
  (6 inches) in diameter
250 g (9 oz) strawberries, hulled
1 teaspoon caster (superfine) sugar
250 ml (9 fl oz/1 cup) thick (double/
  heavy) cream

Using your hands, break the meringues into pieces.

Cut the strawberries into quarters and combine them in a large bowl with the sugar. Using a potato masher or the back of a spoon, squash the strawberries slightly so they start to release their juices.

Whip the cream until soft peaks form, then add to the strawberries in the bowl. Add the meringue and gently stir to just combine.

Spoon into chilled glasses and serve immediately.

CREME ANGLAISE
MAKES ABOUT 350 ML (13 FL OZ); SERVES 4–6

1 vanilla bean, split lengthways
300 ml (10½ fl oz) milk
3 egg yolks, lightly beaten
2½ tablespoons caster (superfine) sugar

Scrape the seeds from the vanilla bean then combine the seeds and bean with the milk in a saucepan. Bring nearly to a simmer then remove from the heat.

Combine the egg yolks and sugar in a bowl then, using electric beaters, whisk until thick and pale. Strain the milk over the egg mixture, stirring to combine well.

Clean the saucepan, return the mixture to it and cook, stirring constantly, over medium-low heat, until thick enough to coat the back of a wooden spoon. Do not let the mixture boil, or the custard will curdle. Strain the custard into a clean bowl, lay plastic wrap directly on the surface to prevent a skin from forming, and allow to cool. The custard will keep, refrigerated, for up to 2 days. Serve with poached fruits, or fruit or chocolate desserts.

Petits pots de crème
SERVES 4

1 vanilla bean
400 ml (14 fl oz) milk
3 egg yolks
1 egg, lightly beaten
80 g (2³/₄ oz/¹/₃ cup) caster (superfine)
   sugar

Preheat the oven to 140°C (275°F/Gas 1).

Split the vanilla bean lengthways, scrape out the seeds then combine the bean and seeds with the milk in a saucepan. Bring the milk just to the boil.

Meanwhile, whisk together the egg yolks, egg and sugar in a bowl until well combined. Strain the milk over the egg mixture and stir to combine. Skim the surface to remove any foam.

Ladle the mixture into four 125 ml (4 fl oz/¹/₂ cup) ramekins and place in a roasting tin. Pour enough hot water into the tin to come halfway up the sides of the ramekins. Bake for 30 minutes, or until the custards are just firm to the touch. Transfer the ramekins to a wire rack to cool, then refrigerate until ready to serve.

Coconut ice
MAKES 30 PIECES

280 g (10 oz/2¹/₄ cups) icing
   (confectioners') sugar
¹/₄ teaspoon cream of tartar
1 egg white, lightly beaten
80 ml (2¹/₂ fl oz/¹/₃ cup) condensed milk
155 g (5¹/₂ oz/1³/₄ cups) desiccated
   (grated dried) coconut
pink food colouring

Brush a 26 x 8 x 4.5 cm (10¹/₂ x 3¹/₄ x 1³/₄ inch) tin with oil and line the base with baking paper.

Sift the icing sugar and cream of tartar into a mixing bowl and make a well in the centre.

Combine the egg white and condensed milk in a small bowl, stirring to mix well, then add to the well in the icing sugar mixture with half the coconut. Using a wooden spoon, stir to combine, then add the remaining coconut and, using your hands, mix to combine well.

Divide the mixture into two bowls and tint one bowlful pink with a few drops of colouring. Knead the colour evenly through, then press the pink mixture evenly over the base of the tin. Press the white mixture over, smoothing the surface.

Refrigerate for 1 hour, or until firm, then cut into squares. Coconut ice will keep, stored in an airtight container in a dark place, for up to 2 weeks.

ORANGE AND APRICOT RICE CAKE

SERVES 8

APRICOT FILLING
200 g (7 oz) dried apricots
115 g (4 oz/½ cup) caster (superfine)
  sugar
125 ml (4 fl oz/½ cup) sweet sherry

1 vanilla bean or 1 teaspoon natural
  vanilla extract
200 g (7 oz/1 cup) medium-grain rice
1 litre (35 fl oz/4 cups) milk
1 fresh bay leaf, bruised
2½ teaspoons finely grated orange zest
4 eggs, lightly beaten
170 g (6 oz/¾ cup) caster (superfine)
  sugar
200 g (7 oz) fresh ricotta cheese (see Note)
60 g (2¼ oz/½ cup) slivered almonds
icing (confectioners') sugar, for dusting
whipped cream, to serve (optional)

To make the filling, combine the apricots in a heatproof bowl with 625 ml (21½ fl oz/2½ cups) boiling water, cover, then stand for 1 hour or until softened. Combine the apricot mixture in a saucepan with the sugar and sherry and slowly bring to the boil. Reduce the heat to low and simmer for 20 minutes, or until the apricots are very soft and pulpy. Cool the apricots in the liquid, then strain, reserving the liquid to use as a syrup when serving.

If using the vanilla bean, split the bean lengthways, scrape out the seeds, then combine the bean and seeds in a saucepan with the rice, milk and bay leaf. Bring slowly to a simmer, then cover and cook, without stirring, over medium-low heat for 15–20 minutes, or until the rice is tender and most of the liquid has been absorbed. Cover and allow to stand for 20 minutes, or until cooled slightly. Remove the bay leaf and vanilla bean.

Preheat the oven to 170°C (325°F/Gas 3). Lightly grease a 20 cm (8 inch) round springform cake tin and line the base with baking paper. Wrap a piece of foil tightly around the base and up the outside of the tin to completely seal it.

Combine the orange zest, eggs, sugar, ricotta (and vanilla extract, if using) in a bowl and, using a wooden spoon, stir until smooth. Add the egg mixture to the rice and stir to combine well. Pour half of the rice mixture into the prepared tin and smooth the top. Arrange the apricots on top, pour over the remaining mixture and scatter over the almonds.

Put the cake tin in a roasting tin and pour in enough boiling water to come halfway up the side of the cake tin. Bake for 50 minutes, or until the cake is firm in the centre. Cover the cake with foil halfway through cooking if it browns too quickly.

Cool the rice cake in the tin. Turn out onto a serving plate, dust with icing sugar and serve with whipped cream and the reserved syrup.

NOTE: Buy ricotta in bulk from a delicatessen or cheese shop; it has a better texture and fresher flavour than the bland, paste-like ricotta sold in tubs in supermarkets.

Cinnamon ice cream
SERVES 6

1 litre (35 fl oz/4 cups) milk
2 wide pieces lemon zest, about 3 cm
  (1¼ inches) long
3 cinnamon sticks
375 g (13 oz/1⅔ cups) caster (superfine)
  sugar
6 egg yolks, lightly beaten
1 teaspoon ground cinnamon

Combine the milk, zest, cinnamon sticks and half the sugar in a saucepan and heat until nearly coming to the boil, then remove from heat, cover and allow to stand for 10 minutes.

Combine the yolks, remaining sugar and ground cinnamon in a bowl and, using electric beaters, whisk until the mixture is thick and pale. Strain the milk mixture, discarding the solids, then pour the milk over the egg mixture, stirring constantly to combine well.

Clean the saucepan, return the mixture to it and cook over low heat, stirring constantly with a wooden spoon, for 5–10 minutes, or until thickened enough to coat the back of the spoon. Take care not to overheat the custard or it will curdle. Strain the custard into a bowl, press plastic wrap onto the surface of the custard to prevent a skin from forming, then allow to stand at room temperature until cold.

When cold, churn the mixture in an ice-cream machine according to the manufacturer's instructions. Alternatively, pour it into a 1.25 litre (44 fl oz/5 cup) shallow metal container. Freeze for 2 hours, or until firm. When half frozen around the edges, beat well then freeze again. Repeat the beating and freezing process twice more.

Key lime pie
Serves 8

125 g (4¹/₂ oz) sweet wheatmeal biscuits, broken
90 g (3¹/₄ oz/¹/₃ cup) butter, melted
4 egg yolks
400 g (14 oz) tin sweetened condensed milk
125 ml (4 fl oz/¹/₂ cup) lime juice
2 teaspoons finely grated lime zest, plus extra to garnish
250 ml (9 fl oz/1 cup) pouring (whipping) cream, to serve

Using a food processor, process the biscuits until finely crushed then transfer to a bowl, add the butter and mix to combine thoroughly. Press into a 23 cm (9 inch) round dish or cake tin and refrigerate until firm.

Preheat the oven to 180°C (350°F/Gas 4).

Using electric beaters, beat the yolks, condensed milk, lime juice and zest for 1 minute, or until smooth and well combined. Pour into the crust and smooth the surface. Bake for 20–25 minutes, or until set, then transfer to a wire rack and allow to cool.

Refrigerate the pie for 2 hours, or until well chilled. Decorate with whipped cream and the extra lime zest just before serving.

ALMOND CROISSANT PUDDING
SERVES 6

4 good-quality, day-old almond croissants,
  torn into small pieces
4 eggs
90 g (3¼ oz) caster (superfine) sugar
250 ml (9 fl oz/1 cup) milk
250 ml (9 fl oz/1 cup) pouring (whipping)
  cream
½ teaspoon finely grated orange zest
80 ml (2½ fl oz/⅓ cup) orange juice
2 tablespoons flaked almonds
icing (confectioners') sugar, for dusting
pouring (whipping) cream or ice-cream,
  to serve

Preheat the oven to 180°C (350°F/Gas 4). Grease the base and side of a 20 cm (8 inch) deep-sided cake tin and line the base with baking paper. Place the croissant pieces in the tin.

Using electric beaters, beat the eggs and sugar together until thick and pale.

Combine the milk and cream in a saucepan and bring almost to the boil. Gradually pour the milk mixture over the egg mixture, stirring constantly. Add the orange zest and juice and stir well. Slowly pour the mixture over the croissants, allowing the liquid to be absorbed before adding any more. Sprinkle over the flaked almonds then bake for 45 minutes, or until cooked when tested with a skewer.

Cool the pudding in the tin for 10 minutes, then invert onto a serving plate. Cut the pudding into wedges, dust with icing sugar and serve warm with cream or ice-cream.

CLASSIC SHORTBREAD
MAKES 16 WEDGES

225 g (8 oz) unsalted butter
115 g (4 oz/^{1}/$_2$ cup) caster (superfine)
  sugar, plus extra for dusting
225 g (8 oz/1^3/$_4$ cups) plain (all-purpose)
  flour
115 g (4 oz/^{2}/$_3$ cup) rice flour

Preheat the oven to 190°C (375°F/Gas 5). Lightly grease two baking trays.

Cream the butter and sugar in a bowl using electric beaters until pale and fluffy. Sift in the flour, rice flour and a pinch of salt and, using a wooden spoon, stir into the creamed mixture until it resembles fine breadcrumbs. Transfer to a lightly floured work surface and knead gently to form a soft dough. Cover with plastic wrap and refrigerate for 30 minutes.

Divide the dough in half and roll out one half on a lightly floured surface to form a 20 cm (8 inch) round. Carefully transfer to a prepared tray. Using a sharp knife, score the dough into eight equal wedges, prick the surface lightly with a fork, and using your fingers, press the edges to create a fluted effect. Repeat this process using the remaining dough to make a second round. Lightly dust the shortbreads with the extra sugar.

Bake for 18–20 minutes, or until the shortbreads are light golden. Remove from the oven and, while still hot, follow the score marks and cut into wedges. Cool on the baking tray for 5 minutes, then transfer to a wire rack.

The shortbread will keep, stored in an airtight container, for up to 1 week.

NOTE Shortbread can be made with plain flour alone; however, the addition of rice flour produces a lighter result.

INDIAN RICE PUDDING
SERVES 6

155 g (5½ oz/¾ cup) basmati rice
20 cardamom pods
2.5 litres (88 fl oz/10 cups) milk
30 g (1 oz/⅓ cup) flaked almonds
175 g (6 oz/¾ cup) sugar
30 g (1 oz/¼ cup) sultanas (golden raisins)

Wash the rice, soak for 30 minutes in cold water, then drain well.

Remove the seeds from the cardamom pods and lightly crush them using a spice grinder or a mortar and pestle.

Bring the milk to the boil in a large heavy-based saucepan and add the rice and cardamom. Reduce the heat to low and simmer for 1½–2 hours, or until the mixture is very creamy, stirring occasionally to prevent the rice from sticking to the pan.

Dry-fry the almonds in a non-stick frying pan over medium-low heat for a few minutes. Reserve a few of the almonds and sultanas for garnish, then add the remainder to the rice mixture and stir the rice mixture to combine well. Divide among bowls and serve warm, garnished with the reserved almonds and sultanas.

CARAMEL SQUARES
MAKES 15 SQUARES

185 g (6½ oz/1½ cups) plain (all-purpose) flour
1½ tablespoons caster (superfine) sugar
100 g (3½ oz) butter, chopped
1 egg

CARAMEL
400 g (14 oz) tin sweetened condensed milk
20 g (¾ oz) butter
1 tablespoon golden syrup, dark corn syrup or treacle

CHOCOLATE TOPPING
120 g (4¼ oz) dark chocolate, chopped
40 g (1½ oz) butter

Brush a 17 x 26 cm (6½ x 10½ inch) shallow rectangular tin with melted butter and line the base with baking paper.

Combine the flour and sugar in a bowl, add the butter then, using your fingertips, rub the butter in until the mixture resembles breadcrumbs. Stir in the egg and enough water to form a coarse dough, then press the dough together on a lightly floured surface. Wrap the dough in plastic wrap and refrigerate for 30 minutes, or until firm.

Preheat the oven to 210°C (415°F/Gas 6–7). Roll out the pastry between two pieces of baking paper to fit the base of the tin. Cover the pastry in the tin with baking paper, fill with baking beads, dried beans or rice then bake for 10 minutes. Remove the beads and paper then bake for another 10 minutes or until golden. Remove from the oven.

Reduce the oven to 180°C (350°F/Gas 4).

To make the caramel, combine all the ingredients in a small saucepan then stir constantly over medium-low heat until the butter melts and the mixture boils and thickens. Spread the caramel in a thin even layer over the pastry and bake for 10 minutes, or until firm. Allow to cool.

To make the topping, combine the chocolate and butter in a bowl set over a saucepan of simmering water. Stir the mixture until melted and well combined, allow to cool slightly then pour over the caramel. Leave to set, then cut into squares.

The squares will keep, stored in an airtight container in a cool, dark place, for 2 days.

Passionfruit tart
Serves 8

135 g (4³/₄ oz) plain (all-purpose) flour
3 tablespoons icing (confectioners') sugar
3 tablespoons custard powder or instant
  vanilla pudding mix
45 g (1¹/₂ oz) butter, cut into cubes
4 tablespoons light evaporated milk

Filling
125 g (4 oz/¹/₂ cup) ricotta cheese (see
  Note, page 60)
1 teaspoon natural vanilla extract
30 g (1 oz/¹/₄ cup) icing (confectioners')
  sugar
2 eggs, lightly beaten
4 tablespoons passionfruit pulp (about
  4 passionfruit)
185 ml (6 fl oz/³/₄ cup) light evaporated
  milk
icing (confectioners') sugar, for dusting

Preheat the oven to 200°C (400°F/ Gas 6). Lightly grease a
23 cm (8¹/₂ inch) loose-based flan (tart) tin. Sift the flour, icing
sugar and custard powder into a bowl and rub in the butter
until crumbs form. Add enough evaporated milk to form a
soft dough. Turn out on a floured surface, bring the dough
together ,then gather into a ball, wrap in plastic wrap and chill
for 15 minutes.

Roll the pastry out on a lightly floured surface to fit the
tin, trim, then refrigerate for 15 minutes. Cover with baking
paper and fill with baking beads, dried beans or rice. Bake for
10 minutes, remove the rice or beans and paper and bake for
another 5–8 minutes, or until golden. Allow to cool.

Reduce the oven to 160°C (315°F/Gas 2–3).

Beat the ricotta with the vanilla extract and icing sugar
until smooth. Add the eggs, passionfruit pulp and evaporated
milk, then beat well. Put the tin with the pastry case on a
baking tray and pour in the filling. Bake for 40 minutes, or
until set. Cool in the tin. Dust with icing sugar to serve.

cultured milk

Fermented milk products have nourished populations the world over for millennia. It is thought that the use of fermented milk dates from about 10,000 BC and that the process was probably first activated spontaneously, when milk stored in goat-skin bags, slung over the backs of camels in African deserts, began to ferment.

Such a climate provided ideal conditions for lactic bacteria to flourish in the fresh milk. Fermented milk products eventually spread widely throughout Europe and the Middle East. Some, such as yoghurt, buttermilk and sour cream, are widely known, but there are also many regional variations of fermented milk products.

The main advantage of fermented milk was that it extended the life of fresh milk. The main bacteria used in fermentation, lactolocci and lactobacilli, are natural inhibitors of the pathogens that cause milk to spoil. Fermentation also enhances the milk's flavour, and the bacilli produced are believed to have healthful properties, aiding in the digestion of the milk. These days fermentation is carried out under highly controlled conditions, but our ancestors relied on far more hit-and-miss procedures. Technically, cheese is also a fermented milk product, but yoghurt, buttermilk, sour cream and crème fraîche are those products most associated with the term 'cultured milk'.

Although fermentation is a fairly straightforward process, numerous factors contributed to the development of such a wide range of products, including the chemical composition of the milk, the additives and starter cultures used, and the way the product is processed. These affect the flavour, texture and consistency of the final product. In factory-made products, it is standard practice for pectins and gums to be used; these stabilisers stop milk solids from forming sediment and also halt any separation of the whey.

These days, carefully measured doses of starter cultures are added to the milk; these differ depending on the product being made. Essentially these cultures are micro-organisms, and they give specific characteristics to the finished fermented milk product. Lactic acid starters ferment lactose in the milk into lactic acid, contributing flavour, aroma and alcohol production as they do so, as well as inhibiting the growth of micro-organisms that cause spoilage. A single strain of bacteria may be added, or a mixture of several; these all start acting at different temperatures.

As the starter cultures grow within the milk, fermentation takes place; this is essentially the chemical conversion of carbohydrates into alcohols or acids. In fermented milk products, both alcohol and lactic acid may be produced, or just lactic acid. The bacteria ingest the milk sugar (lactose), and produce lactic acid as waste material, which increases the acidity levels. This rise in acidity causes the milk proteins to first unfold and then tangle themselves into curds; it also kills off organisms that are not acid tolerant. Following the completion of fermentation, flavourings are sometimes added.

SOUR CREAM

Sour cream has been around for a very long time — some sources suggest since about 10,000 BC. Sour cream is cream that has fermented, with the bacteria that enters the cream causing lactic acid to form, thus thickening and flavouring the cream. These days, as with nearly all modern dairy foods, this process is achieved by introducing a culture to cream; pasteurising stops the fermetation process. Commercially made sour cream has about 18–20 per cent fat; a low-fat variety, made from half milk and half cream, carries about 40 per cent less fat. There is also a non-fat type. These lower-fat types get their texture from the addition of gelatine and vegetable enzymes (although many full-fat commercial sour creams are also subject to these additives).

Sour cream is perhaps most synonymous with the cooking of Eastern European nations such as Hungary and parts of Russia, although over the past 50 years its popularity has spread across the world. It can be used as a topping for soups and hot vegetables, incorporated into cold sauces, soups and salad dressings (it has far fewer calories than mayonnaise and can be used in its place) and baked into quiche fillings and pies. It is used in dips and spreads as well as any number of desserts, both hot and cold. Some of the dishes it is most famously associated with are beef stroganoff and veal goulash. Unless there is flour mixed into it, sour cream will separate if brought to too high a temperature or boiled, so in such cases, bring it to room temperature first, then stir some of the hot liquid into the cream before adding all the cream to the dish. Add it at the last moment, before serving, and if you must return the dish to the heat, do so just long enough to heat the dish, without allowing the sauce to boil. Sour cream can be stabilised by adding 2 tablespoons of plain (all-purpose) flour to each cup of cream. An unopened container of sour cream will keep for about four weeks (check the expiry date) but once opened, should be used within seven days. Discard sour cream if it develops green or pink patches, as these indicate mould.

CRÈME FRAÎCHE

Crème fraîche, which literally means 'fresh cream', originated in France. A fermented cream product, it was originally made by allowing unpasteurised cream to naturally turn sour at warm room temperature, during which time it would thicken and take on a pleasantly tangy flavour. These days crème fraîche is made using pasteurised cream to which a culture is added. It is left to sour and thicken at room temperature for about 24 hours; you can make it at home by adding 2 tablespoons of buttermilk to 1 cup of fresh cream and heating it to 40°C (105°F). Leave it to stand in a draught-free place for 24–36 hours to thicken and ripen, stirring it occasionally. The 'good' bacteria in the buttermilk will protect the cream from spoiling. Crème fraîche keeps for up to 10 days in the refrigerator and will continue to thicken and develop flavour during this time; discard it if it starts to grow mould.

Crème fraîche is prized for its smooth, rich flavour and sumptuous texture. It has distinct advantages over sour cream in that it has double the shelf life, can be whipped like fresh cream and can be heated without separating; it also tastes somewhat sweeter. It has a butterfat content of about 30 per cent, although this varies from brand to brand, as does its consistency. Some brands of crème fraîche are thick and spoonable, while others are as firm as room-temperature margarine. Crème fraîche should be stored, tightly sealed, in the refrigerator. Stir it into hot sauces and soups for a touch of tangy richness or spoon it over poached fruits for dessert. Whipped, it makes a lovely accompaniment to hot desserts such as fruit crumbles, cobblers and crisps, fruit pies and steamed puddings. It can essentially be used in place of both sour cream (for example, with caviar and blinis) and fresh cream. It takes well to slight sweetening (but not too much, as its slightly sour flavour is part of its allure) and to flavouring with chopped fresh herbs such as dill, tarragon, oregano and chives.

YOGHURT

Yoghurt is thought to have originated in Mesopotamia some 3,500 years ago and, like many foods made by processes of natural fermentation, it is likely that it was first made by accident when milk became infected with certain bacteria, then thickened as a result of this. Over time, people worked out a way to do this deliberately and yoghurt became a very important part of the diet of people in many countries throughout the world.

Yoghurt can vary greatly in consistency and flavour, but all yoghurt is made in essentially the same way, and mostly from cow's milk — although yak, buffalo, goat and sheep milk versions are also popular in various countries. To make yoghurt, milk is heated to 93°C (200°F) and kept at that temperature for about 30 minutes (the longer it is heated, the thicker the final yogurt) then rapidly cooled to 44°C (112°F) and mixed with a bacterium to act as a yoghurt starter. It is then incubated at a warm temperature of 37°C (100°F) for about 4 hours — the longer it incubates, the sourer the finished yoghurt. To stop the incubation process, the mixture is put in the refrigerator, where it will last for 7–10 days. The fermentation caused by the bacteria produces lactic acid. The acid acts on the milk proteins so they form a solid, silky mass; acids are responsible for yoghurt's tangy flavour while also creating an environment hostile to dangerous bacteria. The most commonly used bacteria are *Streptoccocus thermophilus* and *Lactobacillus bulgaricus*, although other popular ones are *L. acidophilus* and *L. casei*. These last three are all thought to have many health-giving properties, including promoting good digestion and protecting the body from parasites and other harmful organisms. They are thought to boost the immune system as well.

Flavoured, sweetened yoghurts have helped popularise this food in such places as America, Canada, the United Kingdom and Australia over the last 40 years, before which it was virtually unknown. Many commercially prepared yoghurts have gelatine or pectin added to thicken and stabilise them, and many flavoured low-fat yoghurts are very high in added sugar. Full-fat yoghurt contains 6–10 per cent fat, medium-fat yoghurt 3–5 per cent, and low-fat yoghurt 1–3 per cent. Lastly, there is yoghurt that is made entirely from skimmed milk, which contains no fat at all.

Yoghurts are either set in the tubs they are sold in or stirred before packaging, often with fruit or other flavours stirred in. This type of yoghurt is often also called 'Swiss-style'. Strained yoghurt has had the whey and excess water strained out; the super-thick Greek- and Bulgarian-style yoghurts, with their full, smooth texture and stronger, sharper flavours, are examples of strained yoghurt. These tend to be highest in fat. Some luxury-brand yoghurts have cream added to them to give extra richness.

Yoghurt is used commonly on its own, either chilled or at room temperature, or combined with other chilled or room temperature ingredients for use in chilled soups, dips, dressings and fresh fruit desserts, for example. Sometimes it is used in cake batters and bread doughs and on occasion is stirred into curries and other hot preparations at the very last moment. Yoghurt requires stabilising if it is to be added to simmering liquids such as soups, curries, stews and the like, as it separates when heated. To stablise yoghurt, combine 1 tablespoon plain (all-purpose) flour or cornflour (cornstarch) with just enough water to make a thick, smooth paste and stir this into 500 g (1 lb 2 oz/2 cups) yoghurt before using it in cooking. Note that heating yoghurt destroys the bacterial culture. The yoghurt must be simmered for about 5 minutes to thicken and cook the added starch.

BUTTERMILK

Traditionally, buttermilk is a by-product of butter churning. Also known as 'churn buttermilk', this is a slightly sour liquid left over after butter is made. Its use dates from the time before widespread industrialisation, when people made their own butter and wasted nothing in the process. Nowadays, mass-produced buttermilk is simply milk (either low-fat or skim) that has had a lactic acid bacterial culture added to it and been left to ferment at a low temperature for about 12 hours. It is made from pasteurised milk and its flavour depends upon the types of cultures used to sour it. These days the liquid left after industrial butter making is dried or reduced, then used by the baking industry.

Cultured buttermilk is thicker and more tart than churn buttermilk. It is lower in fat than regular milk, with a fat content of about 2 per cent, and is suitable for those who are sensitive to lactose as, due to the bacteria, it is easier for them to digest than regular milk. Because it is a cultured product, it keeps longer under refrigeration than regular, fresh milk — up to 10 days.

These days, the main use for cultured buttermilk is in baking. Because of its acid content, buttermilk reacts with chemical raising agents to give a very light result, and a recipe requires less baking powder and/or bicarbonate of soda (baking soda) when using buttermilk. Some sources recommend using only half the amount of leavening if you are substituting buttermilk for regular milk in a recipe, although each recipe will vary slightly and you may have to do a little experimenting to convert your favourite baking recipes for using buttermilk.

Buttermilk can also be used as the liquid in mashed potatoes, can form the base for low-fat salad dressings, and can even be made into cold desserts such as jelly, ice cream and bavarois.

Lamb kebabs with mint buttermilk sauce
Serves 4

5 garlic cloves, chopped
5 cm (2 inch) piece ginger, chopped
3 green chillies, chopped
1 onion, chopped
3 tablespoons Greek-style yoghurt
3 tablespoons coriander (cilantro) leaves
500 g (1 lb 2 oz) minced (ground) lamb
red onion rings and lemon wedges, to
    serve

MINT BUTTERMILK SAUCE

1 teaspoon cumin seeds
1/2 cup mint leaves, chopped
1/2 cup coriander (cilantro), chopped
2 cm (3/4 inch) piece ginger, chopped
2 green chillies, chopped
310 ml (10¾ fl oz/1¼ cups) Greek-style
    yoghurt
310 ml (10¾ fl oz/1¼ cups) buttermilk
sea salt

Combine the garlic, ginger, chilli, onion, yoghurt and coriander leaves in a food processor and process until a thick, smooth paste forms. Add the pepper, season to taste with salt then combine with the lamb in a bowl and mix well.

Divide the mixture into 16 even-sized pieces, about 2 tablespoons each, then shape into an oval patty, cover and chill for 20 minutes.

Meanwhile, make the mint buttermilk sauce. Place a small, heavy-based frying pan over low heat, add the cumin seeds then dry-fry, shaking the pan, for 2 minutes or until aromatic. Cool, then grind the seeds to a fine powder using an electric spice grinder or a mortar and pestle.

Combine the mint, coriander, ginger and chilli in a food processor or blender then process or blend until a smooth paste forms. Alternatively, chop the mixture to a fine paste using a large, sharp knife.

Add the yoghurt and buttermilk to the mixture in the food processor then process until well combined and smooth. Season to taste with sea salt and freshly ground black pepper then stir in the cumin. Heat the grill (broiler) to high. Thread four meatballs onto each of four metal skewers then grill (broil) for about 7 minutes, or until brown. Turn and cook the other side for 6–7 minutes, or until browned and cooked through. Serve with the mint buttermilk sauce, red onion rings and lemon wedges.

POACHED EGGS WITH YOGHURT
SERVES 4

60 g (2¼ oz/¼ cup) butter
1 onion, thinly sliced
250 g (9 oz/1 cup) Greek-style yoghurt
4 large eggs
1 teaspoon hot paprika

Preheat the oven to 150°C (300°F/Gas 2).

Melt 20 g (¾ oz) of the butter in a heavy-based frying pan, add the onion then cook over low heat, stirring often, for 15 minutes, or until golden. Remove from the pan and cool slightly. Combine the onion and yoghurt in a small bowl then season to taste with salt.

Divide the yoghurt mixture among four deep 7.5 cm (3 inch) diameter ovenproof ramekins then place on a tray in the oven to heat gently.

Meanwhile, fill a large, deep frying pan three-quarters full with water, add a pinch of salt and bring to a gentle simmer. Crack an egg into a saucer, then slide the egg into the simmering water. Poach for 3 minutes, then remove carefully with a slotted spoon and pat off any excess water with paper towels. Poach all the eggs in the same way. Place an egg in each ramekin and season with salt and pepper.

Melt the remaining butter in a small saucepan and add the paprika. Drizzle over the eggs and serve at once.

WARM ASPARAGUS WITH CREAMY ORANGE-PEPPER DRESSING
SERVES 4

2 tablespoons extra virgin olive oil
2 tablespoons orange juice
1 teaspoon finely grated orange zest
3 tablespoons crème fraîche
sea salt
20 asparagus spears, trimmed

Preheat a barbecue grill plate or a chargrill pan to medium.

In a small bowl, combine the oil and orange juice, and season well with salt and freshly ground black pepper. Pour half the mixture into a shallow dish for coating the asparagus.

Combine the orange zest and crème fraîche with the remaining juice mixture, then season to taste with sea salt and freshly ground black pepper and set aside.

Put the asparagus in the oil mixture in the dish and toss to coat.

Cook the asparagus on the grill plate, turning often, for 4–5 minutes, or until tender and lightly charred. Transfer to a serving platter, pour the dressing over and serve immediately.

TZATZIKI
MAKES 500 ML (17 FL OZ/2 CUPS)

2 Lebanese cucumbers (about 300 g/ 10½ oz)
400 g (14 oz/1²/₃ cup) Greek-style yoghurt
4 garlic cloves, crushed
3 tablespoons finely chopped fresh mint, plus extra to garnish
1 tablespoon lemon juice

Cut the cucumbers in half lengthways, scoop out the seeds with a teaspoon and discard. Coarsely grate the unpeeled cucumber into a small colander. Sprinkle with a little salt and leave to stand over a large bowl for 15 minutes.

Meanwhile, stir together the yoghurt, garlic, mint and lemon juice in a bowl.

Rinse the cucumber under cold water then, taking small handfuls, squeeze out the excess moisture. Combine the cucumber with the yoghurt mixture and season to taste. Serve immediately or refrigerate until ready to serve. Garnish with the extra mint.

Tzatziki can be served as a dip with flatbread or as a sauce for seafood and meat.

EGGPLANT SALAD WITH GARLIC–YOGHURT DRESSING
SERVES 6

1 kg (2 lb 4 oz) large eggplants
  (aubergines)
125 ml (4 fl oz/½ cup) olive oil
1 onion, finely chopped
½ teaspoon ground cinnamon
4 garlic cloves, crushed
2 x 400 g (14 oz) tins crushed tomatoes
2 tablespoons chopped coriander
  (cilantro)
3 tablespoons chopped flat-leaf (Italian)
  parsley
1 tablespoon lemon juice
2 tablespoons chopped mint
150 g (5½ oz/⅔ cup) Greek-style natural
  yoghurt
25 g (1 oz) toasted pine nuts
toasted country-style bread, to serve

Cut the eggplants into 2 cm (¾ inch) pieces, place in a colander over a bowl and sprinkle generously with salt. Leave for 30 minutes, rinse under cold water, then pat dry with a tea towel (dish towel).

Heat 2 tablespoons oil in a large frying pan and fry the eggplant, in batches, until golden, adding more oil if necessary. Drain well on paper towels.

Heat another 2 tablespoons oil in the pan, add the onion then cook, stirring often, over medium heat for 1 minute. Add the cinnamon and half the garlic, cook for 1 minute, then add the tomatoes. Add the eggplant, bring the mixture to a simmer, then cook over medium-low heat for 1 hour, or until the mixture is quite dry. Add half of both the coriander and parsley. Stir and leave to cool.

Mix 2 tablespoons of oil with the lemon juice and add the remaining crushed garlic and all the mint, then stir into the yoghurt.

Gently toss the pine nuts and remaining fresh herbs through the salad. Serve at room temperature with the toast and garlic–yoghurt dressing.

Vegetable terrine with herb sauce
SERVES 8–10

400 g (14 oz) parsnips, cut into chunks
350 g (12 oz) white sweet potato, cut into
  chunks
8 large silverbeet (Swiss chard) leaves
6 asparagus spears, trimmed
2 small zucchinis (courgettes), trimmed
8 green beans, topped and tailed
250 g (9 oz/1 cup) crème fraîche
2 tablespoons powdered gelatine

HERB SAUCE
1 tablespoon finely chopped parsley
1 tablespoon finely chopped chervil
1 tablespoon finely shredded basil
grated zest of 1 small lemon
300 g (10 ½ oz/1 ⅓ cups) crème fraîche

Cook the parsnip and sweet potato in boiling water for
25 minutes, or until tender, then drain and cool.

Dip the silverbeet leaves in boiling water, then remove
carefully with a slotted spoon and lay flat on tea towels
(dish towels).

Lightly oil a 20 x 7 x 9 cm (8 x 2¾ x 3½ inch) terrine
or loaf tin. Line with a layer of plastic wrap, leaving enough
excess overhanging the sides to cover the top of the terrine.
Line the base and sides of the terrine with the silverbeet leaves,
making sure there are no gaps, and leaving enough hanging
over the sides to cover the top.

Trim the asparagus spears if necessary so they fit the
length of the terrine. Slice each zucchini in half lengthways.
Steam the asparagus, zucchini and beans for 6 minutes, or until
just tender. Drain and refresh briefly in iced water to preserve
their colour. Drain well then pat dry with paper towels.

Purée the parsnip and sweet potato with the crème
fraîche in a food processor, or mash and push through a
sieve, and season well. Put 4 tablespoons of water in a small
bowl and sprinkle with the gelatine. Leave for 5 minutes until
spongy, then put the bowl over a pan of simmering water and
stir until melted. Add to the purée and mix well.

Spoon a quarter of the puréed mixture into the terrine,
then arrange the asparagus spears, top to end, in four rows
on top. Spoon over another quarter of the purée, smoothing
the surface, then arrange the zucchini in two rows, cut sides
upwards, over the purée. Spoon over another quarter of purée,
smoothing the top, then arrange the beans in four rows over
the puree. Spoon over the remaining purée and fold the
overhanging silverbeet leaves and plastic wrap over to cover
the top. Refrigerate overnight. To serve, turn out onto
a platter, peel off the plastic wrap and cut into slices.

To make the herb sauce, stir the finely chopped herbs
and lemon zest into the crème fraîche and season well. Serve
with the vegetable terrine.

CARROT PESTO BAKE
SERVES 4

50 g (1³/₄ oz) butter
60 g (2¹/₄ oz/¹/₂ cup) plain (all-purpose)
    flour
750 ml (26 fl oz/3 cups) milk
160 g (5¹/₂ oz/²/₃ cup) sour cream
100 g (3¹/₂ oz/³/₄ cup) grated cheddar
    cheese, plus 50 g (1³/₄ oz/¹/₃ cup) extra
4 eggs, lightly beaten
2 tablespoons basil pesto
750 g (1 lb 10 oz) carrots, grated
250 g (9 oz or about 15) instant lasagne
    sheets

Grease a 30 x 20 cm (12 x 8 inch) baking dish. Heat the butter in a large saucepan, add the flour and cook, stirring, over low heat for 2–3 minutes, or until the mixture is bubbling. Stirring constantly, slowly add the milk, bringing the mixture to the boil between additions. Stir a little of the hot milk into the sour cream then return this mixture to the pan. Add 1 teaspoon freshly ground black pepper. Stir constantly over medium heat for 5 minutes, or until the mixture boils and thickens. Remove from that heat, stir in the cheese and cool slightly. Gradually add the beaten eggs, stirring constantly.

Preheat the oven to 150°C (300°F/Gas 2).

Pour one-third of the sauce into a separate bowl and reserve. Add the pesto and grated carrot to the remaining sauce, stirring to combine.

Put one-third of the carrot mixture into the dish, then lay a third of the lasagne sheets over the mixture. Repeat this process twice, finishing with a layer of lasagne sheets. Spread the reserved sauce evenly over the top, sprinkle with the extra cheese and bake for 40 minutes.

Remove from the oven and allow to stand for 15 minutes before serving.

CHEESE AND CHIVE CORNBREAD
SERVES 4–6

155 g (5½ oz/1¼ cups) self-raising flour
1 tablespoon caster (superfine) sugar
2 teaspoons baking powder
110 g (3¾ oz/¾ cup) fine polenta
60 g (2¼ oz/½ cup) grated cheddar
   cheese
¼ cup chopped chives
¼ cup chopped flat-leaf (Italian) parsley
2 eggs
250 ml (9 fl oz/1 cup) buttermilk
80 ml (2½ fl oz/⅓ cup) olive oil
butter, for spreading
scrambled or poached eggs and grilled
   bacon, to serve (optional)

Preheat the oven to 180°C (350°F/Gas 4). Lightly grease a 20 x 10 cm (8 x 4 inch) loaf tin.

Sift the flour, sugar, baking powder and 1 teaspoon salt into a bowl. Add the polenta, cheese, herbs, eggs, buttermilk and oil and stir to combine well.

Spoon the mixture into the prepared loaf tin and bake for 45 minutes, or until a skewer inserted into the centre comes out clean. Cool slightly, then turn out onto a wire rack to cool completely. Serve slices of buttered cornbread, warm, cool or toasted, with eggs and bacon, if desired.

SALMON AND POTATO SALAD WITH DILL, BUTTERMILK AND MUSTARD DRESSING
SERVES 6

DILL, BUTTERMILK AND MUSTARD DRESSING
1 garlic clove, crushed
1 1/2 tablespoons dijon mustard
185 ml (6 fl oz/3/4 cup) buttermilk
185 ml (6 fl oz/3/4 cup) whole-egg
  mayonnaise
1/4 cup chopped dill

800 g (1 lb 12 oz) small pontiac potatoes,
  halved or quartered
olive oil, for brushing
800 g (1 lb 12 oz) salmon fillets, skinned
  and bones removed
6 cups watercress sprigs
1 small red onion, very finely sliced

For the dressing, combine all the ingredients in a bowl then whisk until the mixture is smooth. Season to taste with salt and freshly ground black pepper then cover with plastic wrap and refrigerate until needed.

Cook the potatoes in a saucepan of boiling, salted water for 15–20 minutes, or until tender, then drain well and cool.

Heat a frying pan or barbecue chargrill plate over high heat. Brush the salmon all over with oil, season to taste then cook, in batches if necessary, for 1–2 minutes, or until cooked but still a little pink in the middle. Allow the salmon to cool to room temperature, then break into bite-sized pieces. Combine the salmon, potatoes, watercress and onion in a large bowl, season to taste and toss to combine well. Divide among bowls or plates, drizzle with the dressing and serve.

MARINATED YOGHURT CHEESE BALLS
MAKES 18

1.5 kg (3 lb/5 oz) Greek-style yoghurt
2 x 50 cm (20 inch) squares muslin
2 fresh bay leaves
3 sprigs thyme
2 sprigs oregano
500 ml (17 fl oz /2 cups) extra virgin
    olive oil

Place the yoghurt in a bowl with 2 teaspoons salt and mix well. Put the muslin squares one on top of the other and place the yoghurt mixture in the centre. Gather up the corners of the muslin and tie securely with string, then tie the muslin bag securely to a wooden spoon or similar, and suspend over a bowl. Refrigerate and leave to drain for 3 days, or until the yoghurt is the consistency of ricotta cheese. Remove the yoghurt from the cloth and place in a bowl.

Roll tablespoons of mixture into balls and place on a large tray; there should be 18 balls. Cover and refrigerate for 3 hours, or until firm.

Place the balls in a clean, dry 1 litre (32 fl oz/4 cup) glass jar with the bay leaves, thyme and oregano sprigs. Add enough olive oil to fill the jar and cover the balls, seal and refrigerate for up to 1 week. To serve, drain the cheese balls well and bring to room temperature.

NOTE: Yoghurt cheese balls are traditionally served at breakfast or as an appetizer, with olives, bread, cold meats and tomato.

VEGETABLE AND LENTIL SOUP WITH SPICED YOGHURT
SERVES 6

2 tablespoons olive oil
1 small leek, white part only, chopped
2 garlic cloves, crushed
2 teaspoons curry powder
1 teaspoon ground cumin
1 teaspoon garam masala
1 litre (32 fl oz/4 cups) vegetable stock
1 bay leaf
185 g (6½ oz/1 cup) brown lentils
450 g (1 lb) butternut pumpkin (squash), peeled and cut into 1 cm (½ inch) cubes
2 zucchini (courgettes), cut in half lengthways and sliced
400 g (14 oz) tin chopped tomatoes
200 g (7 oz) broccoli, cut into small florets
1 small carrot, diced
80 g (2¾ oz/½ cup) peas
1 tablespoon chopped mint, to garnish

SPICED YOGHURT
250 g (9 oz/1 cup) Greek-style yoghurt
1 tablespoon chopped coriander (cilantro) leaves
1 garlic clove, crushed
3 dashes Tabasco sauce

Heat the oil in a saucepan over medium heat. Add the leek and garlic and cook for 4–5 minutes, or until soft and light golden. Add the curry powder, cumin and garam masala and cook for 1 minute, or until fragrant.

Add the stock, bay leaf, lentils and pumpkin. Bring to the boil, reduce the heat to low and simmer for 10–15 minutes, or until the lentils are tender. Season well.

Add the zucchini, tomato, broccoli, carrot and 500 ml (17 fl oz/2 cups) water and simmer for 10 minutes, or until the vegetables are tender. Add the peas and simmer for 2–3 minutes, or until tender.

To make the spiced yoghurt, place the yoghurt, coriander, garlic and Tabasco in a small bowl and stir to combine well. Divide the soup among warmed bowls, top each with a spoonful of spiced yoghurt, scatter with the mint and serve immediately.

ROASTED FIELD MUSHROOMS WITH TARRAGON AND LEMON CRÈME FRAÎCHE
SERVES 4

80 ml (2½ fl oz/⅓ cup) olive oil
2 tablespoons lemon juice
4 garlic cloves, crushed
12 large flat field mushrooms, cleaned and
  stems trimmed
2 tablespoons finely chopped flat-leaf
  (Italian) parsley
toasted sourdough bread, to serve

TARRAGON AND LEMON CRÈME FRAÎCHE
60 ml (2¼ oz/¼ cup) crème fraîche
2 teaspoons lemon juice
1 garlic clove, crushed
2 teaspoons chopped tarragon

Preheat the oven to 200°C (400°F/ Gas 6).

Combine the oil, lemon juice and garlic in a large roasting tin, add the mushrooms, and toss to coat well. Season to taste with salt and pepper and arrange in a single layer in the tin. Roast for 30 minutes, turning occasionally so that the mushrooms cook evenly.

Meanwhile, in a small bowl, combine the crème fraîche, lemon juice, garlic and tarragon and stir until smooth.

Divide the mushrooms among warmed plates and spoon over the cooking juices. Sprinkle with parsley. Serve with the lemon crème fraîche drizzled over and toasted sourdough.

BORLOTTI BEAN MOUSSAKA

SERVES 6

250 g (9 oz/1¼ cups) dried borlotti beans
2 large eggplants (aubergines)
80 ml (2½ fl oz/⅓ cup) olive oil
1 onion, chopped
1 garlic clove, crushed
125 g (4½ oz) button mushrooms, wiped
  clean and sliced
250 ml (9 fl oz/1 cup) red wine
2 x 440 g (15½ oz) tins diced, peeled
  tomatoes
1 tablespoon tomato paste (concentrated
  purée)
1 tablespoon chopped oregano

TOPPING
250 g (9 oz/1 cup) plain yoghurt
4 eggs, lightly beaten
500 ml (17 fl oz/ 2 cups) milk
¼ teaspoon sweet paprika
50 g (1¾ oz/½ cup) grated parmesan
  cheese
40 g (1½ oz/½ cup) fresh breadcrumbs

Soak the borlotti beans in cold water overnight. Drain well and rinse. Transfer the beans to a saucepan, cover with cold water and bring to the boil. Reduce the heat to a simmer and cook the beans over low heat for 1½ hours, or until tender. Drain well.

Meanwhile, preheat the oven grill (broiler) to medium-high. Slice the eggplant, sprinkle with salt and allow to stand for 30 minutes.

Rinse the eggplant then pat dry on paper towels. Brush the eggplant all over with oil, then grill for 3 minutes on each side, or until golden. Drain the eggplant on paper towels.

Preheat the oven to 200°C (400°F/Gas 6).

Heat the remaining oil in a large, heavy-based saucepan. Add the onion and garlic and cook over medium heat for 4–5 minutes, or until the onion is golden. Add the mushrooms and cook for 3 minutes, or until lightly browned. Add the wine and cook over high heat for 2–3 minutes. Stir in the tomatoes, tomato paste and oregano. Bring the mixture to the boil then reduce the heat and simmer for 40 minutes, or until the mixture has reduced and thickened.

Spoon the borlotti beans into a large ovenproof dish and top with the tomato sauce and eggplant slices.

To make the topping, whisk together the yoghurt, eggs, milk and paprika, then pour over the mixture in the dish. Allow to stand for 10 minutes.

Combine the parmesan and breadcrumbs in a small bowl then sprinkle over the moussaka. Bake the moussaka for 50–55 minutes, or until hot, bubbling and golden on top.

MUSHROOM PIROSHKI
MAKES 20

310 g (11 oz/2½ cups) plain (all-purpose) flour
180 g (6¼ oz/¾ cup) cold butter, chopped
1 egg yolk
60 g (2¼ oz/¼ cup) sour cream

FILLING
150 g (5½ oz) Swiss brown mushrooms, wiped clean and coarsely chopped
50 g (1¾ oz) butter
1 small onion, finely chopped
95 g (3¼ oz/½ cup) cooked short-grain rice
1 hard-boiled egg, finely chopped
2 tablespoons chopped flat-leaf (Italian) parsley
2 tablespoons finely chopped dill
1 egg, lightly beaten

Sift the flour and ½ teaspoon salt into a large bowl and add the butter. Using your fingertips, rub in the butter until the mixture resembles fine breadcrumbs. Add the combined egg yolk and sour cream then, using a flat-bladed knife, mix until a coarse dough forms, adding a little iced water if necessary.

Turn the dough out onto a lightly floured surface and press together into a smooth ball. Wrap in plastic wrap and refrigerate for 30 minutes.

To make the filling, process the mushrooms in a food processor until finely chopped. Melt the butter in a frying pan, add the onion and cook, stirring often, for 3–4 minutes, or until softened. Add the chopped mushrooms and cook, stirring, for another 3 minutes, then stir in the rice. Transfer the mixture to a bowl and cool. Stir in the chopped eggs and herbs and season to taste with sea salt and freshly ground black pepper.

Preheat the oven the 190°C (375°F/Gas 5).

Cut the pastry in half, then roll each piece out thickly on a lightly floured surface. Using an 8 cm (3¼ inch) plain biscuit (cookie) cutter, cut 10 rounds from each piece of dough. Place 1 tablespoon of filling in the centre of each round. Brush the edge of each round with a little beaten egg then fold over to form a half-moon shape, pinching the edges together to seal. Prick the tops of the pastries several times with a fork then transfer to a baking tray and refrigerate for 30 minutes.

Brush the pastries with beaten egg then bake for 15 minutes, or until golden. Serve hot.

POLENTA AND SOUR CREAM POUND CAKE WITH BOYSENBERRY COMPOTE
SERVES 10–12

150 g (5½ oz) butter
230 g (8½ oz/1 cup) soft brown sugar
115 g (4 oz/½ cup) caster (superfine)
   sugar
5 eggs
185 g (6½ oz/¾ cup) sour cream
½ teaspoon natural almond extract
1 teaspoon natural vanilla extract
155 g (5½ oz/1¼ cups) plain
   (all-purpose) flour
1½ teaspoons baking powder
150 g (5½ oz/1 cup) fine polenta
whipped cream, to serve

BOYSENBERRY COMPOTE
80 g (2¾ oz/⅓ cup) caster (superfine)
   sugar
2 teaspoons lemon juice
500 g (1 lb 2 oz/3¾ cups) boysenberries

Preheat the oven to 180°C (350°F/Gas 4). Grease a 24 x 14 cm (9½ x 5½ inch) loaf tin.

Using electric beaters, cream the butter, brown sugar and caster sugar in a large bowl until pale and fluffy. Add the eggs one at a time, beating well after each addition. Reduce the speed to low and mix in the sour cream and almond and vanilla extracts.

Sift together the flour, baking powder and a pinch of salt. Add the flour mixture and polenta to the butter mixture and gently stir to combine well. Spoon into the prepared tin, smooth the surface, then bake for 50 minutes, or until cooked when tested with a skewer. Cool in the tin for 5 minutes, then turn out onto a wire rack to cool completely.

Meanwhile, make the boysenberry compote. Combine the sugar, lemon juice and 2 tablespoons water in a saucepan, then stir over medium heat for 3 minutes, or until the sugar dissolves. Add the berries, stir to coat, and bring the mixture to a simmer. Cook over medium-low heat for 5 minutes, stirring occasionally, or until the berries are soft but still holding their shape. Cool to room temperature. The compote can be served at room temperature or chilled.

Cut the cake into thick slices and serve toasted with the compote and cream.

CAPPUCCINO AND CHOCOLATE MUFFINS

MAKES 8

20 g (^{3}/$_4$ oz/^{1}/$_4$ cup) instant espresso coffee granules
1 tablespoon boiling water
310 g (11 oz/2^1/$_2$ cups) self-raising flour
115 g (4 oz/^{1}/$_2$ cup) caster (superfine) sugar
2 eggs, lightly beaten
375 ml (13 fl oz/1^1/$_2$ cups) buttermilk
1 teaspoon natural vanilla extract
150 g (5^1/$_2$ oz) butter, melted
100 g (3^1/$_2$ oz/^{2}/$_3$ cup) chopped good-quality dark chocolate
30 g (1 oz) butter, extra
3 tablespoons soft brown sugar

Preheat the oven to 200°C (400°F/Gas 6). Grease the bases of eight 125 ml (4 fl oz/^{1}/$_2$-cup) capacity ramekins. Cut eight rectangular strips of baking paper 8 x 22 cm (3^1/$_4$ x 8^1/$_2$ inches) and roll into cylinders to fit the ramekins. Secure the cylinders with string and place the ramekins on a baking tray.

Dissolve the coffee in the boiling water and cool. Sift the flour and sugar into a large bowl.

Combine the egg, buttermilk, vanilla, melted butter, chocolate and the coffee mixture in a bowl and stir to mix well. Add the buttermilk mixture to the flour mixture then, using a large metal spoon, quickly stir to just combine.

Divide the mixture among the ramekins. Combine the extra butter and brown sugar in a small saucepan and stir over medium heat for 1–2 minutes or until the butter has melted and the sugar dissolves. Spoon the mixture over the muffins and, using a skewer, gently swirl into the top of each.

Bake the muffins for 25–30 minutes, or until cooked when tested with a skewer. Allow to cool slightly in the ramekins before serving.

White chocolate mousse

100 g (3¹/₂ oz/²/₃ cup) chopped good-
  quality white chocolate
125 ml (4 fl oz/¹/₂ cup) milk
2 teaspoons powdered gelatine
400 g (14 oz) fromage fraîs
3 egg whites
3 tablespoons passionfruit pulp
icing (confectioners') sugar, to dust

Place the chocolate and milk in a small saucepan and stir over low heat until the chocolate has melted, then cool. Place 60 ml (¹/₄ cup) cold water in a small heatproof bowl and sprinkle with the gelatine. Leave for 5 minutes until spongy, then put the bowl over a small pan of simmering water and stir until the gelatine has dissolved. Using a wooden spoon, stir the gelatine into the chocolate mixture.

Place the fromage frais in a large bowl and gradually stir in the chocolate mixture, a little at a time, stirring after each addition until smooth.

Beat the egg whites in a clean, dry bowl with electric beaters until soft peaks form. Gently fold the egg whites and the passionfruit pulp into the chocolate mixture. Divide the mixture equally among eight 125 ml (4 fl oz/¹/₂ cup) serving dishes or a 1 litre (32 fl oz/4 cup) glass bowl. Refrigerate for 3 hours, or until set. Serve with a light dusting of icing sugar.

NOTE: It is important to have the ingredients at room temperature to ensure the texture is smooth.

Banana and blueberry pancakes
Makes about 12

250 ml (9 fl oz/1 cup) buttermilk
1 egg, lightly beaten
1 tablespoon melted butter
1 teaspoon natural vanilla extract
115 g (4 oz) plain (all-purpose) flour
1 teaspoon baking powder
2 ripe bananas, mashed
100 g (3½ oz) blueberries
1 teaspoon vegetable oil
maple syrup, to serve

Preheat oven to 120°C (235°F/Gas ½).

Combine the buttermilk, egg, butter and vanilla extract in a bowl and whisk to mix well. Sift in the flour, baking powder and ½ teaspoon salt, then quickly stir until just combined, taking care not to overmix; the batter should be a little lumpy. Stir in the bananas and blueberries.

Heat the oil in a frying pan over medium heat. Working in batches, add 60 ml (2 fl oz/¼ cup) of batter to the pan for each pancake, then cook for 3 minutes, or until bases of pancakes are golden brown. Turn over and cook for another minute, or until risen slightly and cooked through. Cover the cooked pancakes with aluminium foil on a plate and keep warm in the oven while cooking the remaining pancakes. Serve the hot pancakes drizzled with maple syrup.

PINEAPPLE AND SOUR CREAM GRATIN
SERVES 4

800 g (1 lb 12 oz) ripe pineapple, cut into
    1.5 cm (⁵/₈ inch) pieces
60 ml (2 fl oz/¹/₄ cup) dark rum
45 g (1¹/₂ oz) unsalted butter
1 teaspoon natural vanilla extract
¹/₂ teaspoon ground ginger
140 g (5 oz/³/₄ cup lightly packed) soft
    brown sugar
300 g (10¹/₂ oz/1 ¹/₄ cups) sour cream or
    quark
60 ml (2 fl oz/¹/₄ cup) pouring (whipping)
    cream
1 teaspoon finely grated lemon zest

Preheat an oven grill (broiler) to medium-high.

Combine the pineapple, rum, butter, vanilla extract, ginger and ¹/₄ cup of the sugar in a large saucepan. Cook, stirring occasionally, over medium-high heat for 8–10 minutes, or until the sugar has dissolved and the mixture is very reduced and caramelised. Remove from the heat, divide the mixture among four individual gratin dishes and allow to cool slightly.

Combine the sour cream or quark, cream and lemon zest in a bowl and whisk until smooth. Divide the mixture among the dishes, smoothing the tops. Sprinkle the remaining brown sugar over the top, then cook under the grill for 4–5 minutes, or until the sugar has caramelised, taking care not to burn the sugar. Serve immediately.

LEMON FROZEN YOGHURT
SERVES 6–8

1 kg (2 lb 4 oz/4 cups) vanilla-flavoured
    yoghurt
185 ml (6 fl oz/³/₄ cup) lemon juice
185 g (6¹/₂ oz) caster (superfine) sugar
60 ml (2 fl oz/¹/₄ cup) light corn syrup
1 teaspoon finely grated lemon zest
¹/₂ teaspoon natural vanilla extract

Line a fine sieve with cheesecloth (muslin), place over a bowl, add the yoghurt to the sieve and refrigerate for 2 hours, or until the liquid has drained off and the yoghurt has thickened. Discard the liquid.

Combine the remaining ingredients in a bowl and whisk until well combined and the sugar has dissolved. Add the drained yoghurt and whisk in well.

If you have an ice-cream machine, pour the mixture into it and churn according to the manufacturer's instructions. Otherwise, place the mixture in a shallow metal tray and freeze for 2 hours, or until the mixture is frozen around the edges. Transfer to a large bowl and beat until smooth. Repeat this step three times. For the final freezing, place in an airtight container, cover the surface with a piece of greaseproof paper and a lid, and freeze for 4 hours or overnight.

Pineapple and sour cream gratin

soft and fresh cheeses

Fresh cheeses are a special category of cheese, made using simple processes (curdling milk, draining off whey and setting the curd, perhaps with lactic acid). They are designed to retain as much moisture as possible and to preserve and showcase the light, perfumed fragrance and flavour of fresh, best-quality milk.

With their high moisture content (30–60 per cent) and lower salt and fat content (most have between 0.5 and 30 per cent fat), fresh cheeses are highly perishable, and should be consumed within about a week. They are easy to spread and take well to the addition of such ingredients as chopped herbs, olives, garlic and sun-dried tomatoes, so make brilliant dips and sandwich spreads. Some, notably mascarpone, cream cheese and ricotta, are also excellent in desserts, including baked ones such as cheesecakes and pastry fillings. They are not, on the whole, good when heated, as they separate over direct heat; if used in pasta sauces and the like, they need to be added right at the end, and off the heat.

Some fresh cheeses are marinated in oil, pickled in brine or coated in ash to make them last longer. They are made from a variety of milks. Goat's milk fresh cheeses have a slightly zingy, mild, creamy taste marked by a fresh acidity. The ash coating that some have was originally to aid preservation, but is now mainly just cosmetic. Sheep's milk fresh cheeses are rich and deep in flavour and can vary in taste from salty to slightly acidic.

RICOTTA

Ricotta, when made in the traditional way, is technically not a cheese at all as it is made not from milk but from whey, a lactose-rich, liquid by-product of cheese making. Ricotta is Italian, probably originating in Sicily, but Greek sources hint that a ricotta-like food was made as early as the third century AD.

Like many foods, ricotta, which means 'cooked again', was no doubt borne of thrift, as a means of using the whey that would otherwise have been thrown away after making cheese, or fed to pigs. These days ricotta is made on an industrial scale and either whole or skim milk is used. Some whey ricotta is still made, but as very large volumes of whey are required, manufacturers nowadays tend to supplement whey with cow's milk, or make it entirely from milk.

The best ricotta is handmade, from the whey of cow's, water buffalo's or sheep's milk, with some milk added, and has a drier texture and nuttier flavour than mass-produced ricotta. To make ricotta, the whey (or whey and milk mixture) is left to ferment overnight (although these days lactic acid is used to bypass this time-consuming step) then heated to 82°C (180°F) in order to coagulate proteins. These proteins rise to the surface and are skimmed off, cooled and drained; the resulting substance is ricotta.

Snowy-white and moist, ricotta is sweet, mild, milky flavoured and low in salt. It is also low in fat (about 5 per cent). Ricotta is highly perishable, keeping for just 5–6 days in the refrigerator. It should be white; any yellowing or sour smells indicate age or poor storage and the cheese should be discarded. Ricotta can be used in many ways and perhaps some of the most famous ricotta-based dishes are traditional Sicilian desserts, such as cassata and cannoli.

Unsurprisingly, ricotta is most widely used in Italian cooking; it appears in lasagna, on pizzas, and stirred, at the last moment, through pasta. Although it has quite a grainy texture, ricotta becomes quite smooth when beaten with other ingredients, and is popular in cheesecakes, pastry fillings and stuffings for vegetables and ravioli, stirred through frittata mixtures and spread onto bruschetta.

FROMAGE FRAÎS

Fromage fraîs literary means 'fresh cheese'; it originated in Belgium and the north of France and is also called fromage blanc (although fromage blanc tends to contain no salt, whereas fromage fraîs is lightly salted). A fresh cheese, fromage fraîs is made similarly to cottage cheese, where rennet and a starter culture are added to milk. Unlike cottage cheese, fromage fraîs is stirred to give it a characteristic silky-smooth texture, rather like yoghurt.

As such, the term 'fromage fraîs' encompasses many fresh white cheeses made in the same way — some taste slightly tangier than others, depending upon how long they were incubated before draining (some fresh white cheeses are not drained). Fromage fraîs is a low-fat dairy product, although these days many manufacturers add cream to it, increasing the fat levels to about 8 per cent and giving it a richer texture and flavour. Fromage fraîs lends itself to being combined with chopped fresh herbs and used as a dip or spread, or served (either slightly sweetened or not) with desserts, especially fruit ones. It can also be stirred into just-drained pasta (along with, perhaps, chopped sun-dried tomatoes, capers, olives, anchovies, fresh basil or similarly assertive-tasting ingredients) just before serving, for a light, easy sauce. It can also be used in cooked dishes, such as custards or cheesecakes. Fromage fraîs is usually made from cow's milk, but goat's milk versions are also popular. It keeps for up to about 10 days in the refrigerator, but as with all cheeses, always be sure to check expiration dates on the packaging.

QUARK

Quark is a fresh, white cheese, very similar to ricotta. Quark simply means 'curd' in German, and it is thought this type of cheese has been made since the Iron Age. It is a Central European cheese, popular in such countries as Austria and Germany, where it is combined with savoury flavours (such as caraway, paprika or chopped spring onion) and spread on bread, or sweetened with sugar and honey and eaten with desserts or for breakfast. A highly versatile cheese, it is also used in baked preparations such as savoury and sweet pies, mousses and cheesecakes. The texture is firm but a little crumbly and the flavour mild and milky. Quark is made from pasteurised milk which has a starter culture added to it then some rennet to firm the curds. Traditionally the cheese is then hung in cheesecloth (muslin) and left so that some of the whey can drain off. In industrial production, the whey is separated by centrifuge and the quark is packaged in tubs for sale, where it sits in some of its residual whey. Quark contains about 15 per cent fat, although this can vary as it can be made from low-fat, skim or whole milk. Avoid quark with added milk powder, as this makes the texture gritty.

CREAM CHEESE

Cream cheese was invented in New York state in 1872, by a dairy farmer who claimed to have developed a cheese that was richer than any before it. It is believed he was trying to emulate the French soft cheese neufchâtel, and today there is sometimes confusion between these two cheeses. Neufchâtel is, in fact, rather different from cream cheese. It comes from the Normandy region of northern France and has a long history. It has a soft-mould rind like camembert, has been aged for eight weeks or so and is a little salty and sharp. Neufchâtel is made from whole milk, not cream, and has more moisture and less fat than cream cheese.

Cream cheese was commercially distributed from the 1880s, as 'Philadelphia' cream cheese. (At the time, that city was synonymous with top-quality food items, which would be described as being of 'Philadelphia quality'.) The cheese was, and still is, a soft, milky-white cheese made of milk and cream, that has not been allowed to mature and is designed to be consumed fresh. It has a high water content (not more than 55 per cent) and a milk fat content of about 33 percent. It is stabilised with substances such as guar and carob gum to extend its shelf life and give it firmness. These days it is also available in low-fat, non-fat and flavoured versions.

Cream cheese is soft and spreadable at room temperature, making it a popular choice for dips and spreads and for slathering on breads in combination with smoked salmon and other toppings. Perhaps the most common use for cream cheese is in dessert recipes, notably for cheesecakes. Cheesecakes are actually a very old food, being popular among the ancient Greeks (cheesecakes were served to the athletes competing in the original Olympic games) and the Romans, who used them as temple offerings. The Romans were responsible for spreading cheesecakes through Europe as they conquered various nations. The recipe and ingredients for early cheesecake were recorded thus by the Greek writer Athenaeus in about AD 230: 'Take cheese and pound it till smooth and pasty; put cheese in a brazen sieve; add honey and spring wheat flour. Heat in one mass, cool, and serve.'

Although cottage cheese, sour cream, ricotta and neufchâtel are all associated with cheesecakes, today the most famous cheesecakes are the 'New York'-style ones, which are made from little more than cream cheese, cream, sugar and eggs. Cheesecake became popular in New York in the early 1900s, with many restaurants developing their own versions.

COTTAGE CHEESE
Cottage cheese is made when milk (either whole or skim) is heated, then rested, during which time curds form and clump together. The liquid is drained off and because the cheese isn't pressed to remove the liquid, some stays, giving cottage cheese its characteristic moistness. Like ricotta, cottage cheese is white, soft, mild tasting and rather perishable, but unlike ricotta, it has a decidedly lumpy texture and a more salty flavour. The size of the curds varies from brand to brand and also according to how the cheese was made. There are two methods. One involves the addition of rennet, which speeds

the curd-forming process and results in larger curds and cheese with lower acidity. In the other method, no rennet is added; this type has smaller curds and more acid. Creamed cottage cheese has cream added to the curds after rinsing or draining, while low-fat cottage cheese has milk added. The curds may or may not be washed to rid them of excess acidity, hence some brands of cottage cheese taste rather different from others.

Popular with the health conscious and with those on low-fat diets due to its high-protein, low-fat nutritional profile, cottage cheese can be used in dips and spreads, in cheesecakes and sweet pancake fillings, or stirred into hotcake and flapjack mixtures that are destined for the griddle.

MASCARPONE

Mascarpone became the darling of the restaurant scene in the latter decades of the twentieth century. When chefs and diners discovered the classic Italian dessert tiramisu, an ongoing love affair with its chief ingredient, mascarpone, began.

Mascarpone is a fresh cow's milk cheese that originated in an area south-west of Milan, in northern Italy, sometime around the 16th or 17th century. It is thick and spreadable, almost the consistency of room-temperature butter. In colour it is pale ivory and its flavour is rich and sweetly milky. It is used in both sweet and savoury recipes. Theories abound as to how mascarpone got its name — one is that it is from the Spanish for 'better than good' (*mas que bueno*), from the days when the Spanish occupied that part of Italy.

To make mascarpone, cream is skimmed from milk then the cream is heated to 85°C (180°F). Tartaric acid is added, causing the cream to thicken and become quite dense. This mixture is left to stand for about 36 hours so that the whey can drain off; it is not pressed to extract the whey, as many other cheeses are. The mascarpone is then ready to use.

Although it is now available year round, mascarpone was originally seasonal, made in the autumn and winter months. It only keeps for about five days. It is high in energy (about 1885 kj/450 calories per 100 g/3½ oz), with a fat content of 60–75 per cent, and is relatively low in protein.

Mascarpone is arguably at its best when served simply. A little of it stirred through risotto before serving, or spooned on top of polenta, is delicious, as it is when sweetened with a little sugar and served with fresh or poached fruit. Aside from tiramisu, another favoured way in which the Italians enjoy it as a dessert is to serve it plain, with accompanying bowls of cocoa, sugar and perhaps some finely ground espresso, to flavour spoonfuls of mascarpone to taste.

Baked ricotta with ratatouille
Serves 8–10

1.5 kg (3 lb 5 oz) firm ricotta cheese, well drained (see Note, page 60)

4 eggs, lightly beaten

3 garlic cloves, finely chopped

2 tablespoons chopped oregano

sea salt

80 ml (2½ fl oz/⅓ cup) extra virgin olive oil

300 g (10 ½ oz) eggplant (aubergine), cut into 1.5 cm (⅝ inch) pieces

3 capsicums (peppers) (a mixture of yellow, green and red), trimmed, seeded and cut into 1.5 cm (⅝ inch) pieces

400 g (14 oz) tin crushed tomatoes

Preheat the oven to 180°C (350°F/Gas 4) and lightly grease a 22 cm (8½ inch) springform cake tin. Combine the ricotta, eggs, 1 finely chopped garlic clove and 1 tablespoon of the chopped oregano in a bowl and season to taste with sea salt and freshly ground black pepper. Pour the ricotta mixture into the tin then tap the tin twice on a work surface to expel any air bubbles. Bake the ricotta for 1 hour 30 minutes, or until firm and light golden. Cool the ricotta in the pan on a wire rack, pressing down on the ricotta occasionally to remove any air bubbles.

Meanwhile, heat 2 tablespoons of the oil in a frying pan, add the eggplant and cook for 4–5 minutes, or until golden. Add the capsicum and remaining garlic and cook for 5 minutes, until the capsicum softens, adding an extra tablespoon of oil if necessary. Stir in the tomato and remaining oregano and cook for 10–15 minutes, or until the mixture is reduced slightly and the vegetables are tender. Season to taste with sea salt and freshly ground black pepper.

Remove the ricotta from the pan and cut into wedges. Serve the ricotta wedges with some ratatouille on the side.

Pizza rustica
Serves 6

PASTRY

375 g (13 oz/3 cups) plain (all-purpose)
  flour
1 teaspoon icing (confectioners') sugar
155 g (5½ oz/⅔ cup) butter, chilled and
  chopped
1 egg
1 egg yolk
2 tablespoons iced water

FILLING

500 g (1 lb 2 oz/2 cups) ricotta (see Note,
  page 60)
6 eggs, separated
100 g (3½ oz) lean bacon slices, cut into
  small strips
75 g (2½ oz) salami, cut into 5 mm
  (¼ inch) pieces
100 g (3½ oz) grated mozzarella cheese
100 g (3½ oz) smoked mozzarella or
  other naturally smoked cheese, cut into
  1 cm (½ inch) pieces
25 g (¾ oz/¼ cup) grated parmesan
  cheese
1 tablespoon chopped flat-leaf (Italian)
  parsley
½ teaspoon chopped oregano
a pinch of nutmeg
1 egg beaten with 1 tablespoon cold
  water, for glazing

For the pastry, sift the flour, icing sugar and 1 teaspoon salt into a bowl. Rub in the butter with your fingertips until the mixture resembles fine breadcrumbs. Add the egg, yolk and water, 1 teaspoon at a time, cutting it in with a flat-bladed knife until a coarse dough forms. Turn out onto a lightly floured surface and gather together into a smooth ball. Cover with plastic wrap and refrigerate for 30 minutes.

Preheat the oven to 190°C (375°F/Gas 5) and place a baking tray on the centre shelf. Grease a deep 23 cm (9 inch) round pie dish or springform cake tin.

For the filling, place the ricotta in a large bowl and beat until smooth. Gradually add the egg yolks, beating well after each addition. Add the bacon, salami, mozzarella cheeses, parmesan, parsley, oregano and nutmeg. Season well.

Using electric beaters, whisk the egg whites in a large bowl until stiff peaks form, then fold them into the ricotta.

Divide the pastry into two pieces, one slightly larger than the other. Roll the larger piece out on a lightly floured surface to fit the base and sides of the dish. Line the dish and trim the pastry. Roll out the remaining pastry portion to the same thickness for the pie lid. Spread the filling over the base and smooth the surface. Brush the pastry edges with the egg glaze and position the lid on top. Press the edges together firmly to seal, then trim with a sharp knife. Using your fingers, crimp the edges of the pastry, brush the surface well with the egg glaze and prick the surface all over with a fork.

Place the pie dish on the heated tray and bake for 45–50 minutes, or until the pastry is golden and the filling is set. Loosely cover the top with foil if it browns too quickly. Set aside for 20 minutes before serving.

SMOKED SALMON TARTLETS
MAKES 24

250 g (9 oz/1 cup) cream cheese, at room temperature
1 1/2 tablespoons wholegrain mustard
2 teaspoons dijon mustard
2 tablespoons lemon juice
2 tablespoons chopped dill
6 sheets frozen ready-rolled puff pastry, thawed
300 g (10 1/2 oz) smoked salmon, cut into thin strips
2 tablespoons capers, rinsed and drained
dill sprigs, to garnish

Preheat the oven to 210°C (415°F/Gas 6–7). Line two large baking trays with baking paper. Combine the cream cheese, mustards, lemon juice and dill in a food processor. Process until combined and smooth, then transfer to a bowl, cover with plastic wrap and refrigerate.

Cut four 9.5 cm (3 3/4 inch) rounds from each sheet of puff pastry, using a fluted biscuit (cookie) cutter, and place on the baking trays. Prick the pastries all over with a fork then cover with plastic wrap and refrigerate for 10 minutes.

Bake the pastries, one tray at a time, for 7 minutes each, then remove from the oven and, using a spoon, flatten the centre of each pastry. Return to the oven and bake for a further 5 minutes, or until the pastry is golden. Transfer to a wire rack to cool.

Spread some of the cream cheese mixture over each pastry round, leaving a 1 cm (1/2 inch) border. Arrange the salmon over the top. Decorate with a few capers and a sprig of dill. Serve immediately.

HERB-FILLED RAVIOLI WITH SAGE BUTTER
SERVES 4

PASTA
300 g (10½ oz/2½ cups) plain
(all-purpose) flour
3 eggs, beaten
60 ml (2 fl oz/¼ cup) olive oil

FILLING
250 g (9 oz/1 cup) ricotta cheese (see
Note, page 60)
2 tablespoons freshly grated parmesan
cheese, plus extra, shaved, to garnish
2 teaspoons snipped chives
1 tablespoon chopped flat-leaf (Italian)
parsley
2 teaspoons chopped basil
1 teaspoon chopped thyme
sea salt

SAGE BUTTER
200 g (7 oz) butter
24 small sage leaves

Sift the flour into a bowl and make a well in the centre. Gradually mix in the eggs and oil and stir until a coarse dough forms. Turn out onto a lightly floured surface and knead for 6 minutes, or until smooth. Cover with plastic wrap and stand for 30 minutes.

Combine the ricotta, parmesan and herbs in a bowl and season to taste with sea salt and freshly ground black pepper.

Divide the dough into four even-sized pieces. Lightly flour a large work surface and, using a floured rolling pin, roll out one piece, rolling from the centre to the edge. Continue rolling, always rolling from the front of you outwards and rotating the dough often. Fold the dough in half and roll it out again. Repeat the process seven times to make a smooth circle of pasta about 5 mm (¼ inch) thick. Roll this sheet out quickly and smoothly to a thickness of 2.5 mm (⅛ inch). Repeat with the remaining portions to make four sheets of pasta, two slightly larger than the others. Cover with a tea towel (dish towel). Alternatively, roll each portion of dough out using a pasta machine, following the manufacturer's instructions.

Spread one of the smaller sheets on a work surface and place heaped teaspoons of filling at 5 cm (2 inch) intervals. Brush a little water between the mounds of filling; you will be cutting along these lines. Place a larger sheet on top and firmly press the sheets together along the cutting lines. Cut the ravioli out with a pastry wheel or sharp knife and transfer to a lightly floured baking tray. Repeat with the remaining dough and filling.

To make the sage butter, melt the butter over low heat in a small heavy-based saucepan, without stirring or shaking. Carefully pour the clear butter into another pan, discarding any white sediment. Return the clarified butter to the stove, over medium heat. Add the sage leaves and cook until crisp but not brown. Remove and drain on paper towels. Reserve the warm butter.

Cook the ravioli, in batches, in a large saucepan of salted simmering water for 5–6 minutes, or until tender. Top with warm sage butter and leaves and garnish with shaved parmesan.

CHICKEN WITH RICOTTA AND BACON STUFFING

SERVES 6

STUFFING

2 rindless bacon slices, finely chopped
1 tablespoon olive oil
1 small onion, diced
1 garlic clove, crushed
250 g (9 oz/1 cup) ricotta cheese (see
    Note, page 60)
1 egg, beaten
2 tablespoons freshly grated parmesan
    cheese
80 g (2¾ oz/1 cup) fresh breadcrumbs
15 g (½ oz/¼ cup) chopped flat-leaf
    (Italian) parsley
2 tablespoons chopped basil
2 tablespoons chopped oregano
a pinch of freshly grated nutmeg

6 chicken leg quarters
15 g (½ oz) butter, melted

To make the stuffing, heat a non-stick frying pan, add the bacon and cook, stirring often, over medium heat for 6–8 minutes, or until golden. Remove and set aside.

In the same pan, heat the oil, add the onion and garlic and cook over medium heat, stirring often, for 5–6 minutes, or until soft.

Combine the ricotta, egg, parmesan, breadcrumbs, chopped herbs, bacon, onion, garlic and nutmeg in a large bowl and stir to mix well.

Preheat the oven to 180°C (350°F/Gas 4). Using your fingers, ease the skin away from the meat of each piece of chicken to create a pocket, then push the ricotta mixture evenly into the pocket under the skin.

Place the chicken on a lightly oiled baking tray, brush with the melted butter then roast for 45 minutes, or until the chicken is cooked through. Stand the chicken, loosely covered with foil, for 5 minutes, before serving.

DEEP-FRIED ZUCCHINI FLOWERS
SERVES 4

2 eggs
60 g (2¼ oz/½ cup) plain (all-purpose)
  flour
10–12 zucchini (courgette) flowers
oil, for deep-frying

STUFFING
125 g (4½ oz) ricotta cheese
1 tablespoon chopped basil
2 tablespoons grated parmesan cheese
2 tablespoons fresh breadcrumbs
1 egg yolk

Whisk the eggs with the flour and season to taste with salt and pepper.

To make the stuffing for the flowers, combine the ricotta, basil, parmesan, breadcrumbs and egg yolk in a bowl and stir to mix well. Season to taste then use the mixture to stuff the flower cavities.

Half-fill a medium saucepan with oil then heat to 180°C (250°F), or until a cube of bread dropped into the oil turns golden in 15 seconds.

Add a little cold water to the batter if it is too thick, then, working in batches, dip each stuffed zucchini flower into the batter, draining excess batter, and deep-fry for 4-5 minutes or until golden. Drain flowers well on paper towels then serve immediately.

NOTE. Before using zucchini flowers, remove and discard the stamen from inside the flower, wash the flower and make sure it doesn't harbour any insects. They only last a few days after being picked, so buy them on the day you plan to use them.

CERVELLE DE CANUT
SERVES 8

500 g (1 lb 2 oz/2 cups) fromage blanc or
 curd cheese
2 tablespoons olive oil
1 garlic clove, finely chopped
2 tablespoons chopped chervil
4 tablespoons chopped parsley
2 tablespoons snipped chives
1 tablespoon chopped tarragon
4 French shallots, finely chopped
toast or baguette, to serve

Beat the fromage blanc or curd cheese with a wooden spoon,
then add the olive oil and garlic and beat it into the cheese.
Add the herbs and shallots and mix together well. Season to
taste and serve with pieces of toast or bread.

CHIVE AND CHEESE LOG
MAKES A 30 CM (12 INCH) LOG; SERVES 6

500 g (1 lb 2 oz/2 cups) cream cheese,
 softened
1 tablespoon lemon juice
1 garlic clove, crushed
2 teaspoons chopped thyme
2 teaspoons chopped tarragon
1 tablespoon chopped flat-leaf (Italian)
 parsley
60 g (2¼ oz/1 cup) snipped chives
crackers or toast, to serve

Using electric beaters, beat the cream cheese in a large bowl
until soft and creamy, then mix in the lemon juice and garlic.

Combine the thyme, tarragon and chopped parsley in a
small bowl.

Line a 20 x 30 cm (8 x 12 inch) tin with foil, allowing it
to hang over two sides to aid with removal later. Sprinkle the
chives over the base of the tin, then spoon the cream cheese
mixture over the chives. Using a palette knife, gently spread
the mixture evenly into the tin, pushing it into any gaps.
Sprinkle the combined herbs evenly over the top.

Lift the foil from the tin and place on a work surface.
Roll the cheese into a log, starting from the longest edge and
using the foil to help you roll, then cover with plastic wrap
and place on a baking tray.

Refrigerate for at least 3 hours, or overnight. To serve,
remove the log from the foil and place on a platter. Serve with
crackers or toast

POTATO, MINT AND SHEEP'S CHEESE PIES
MAKES 4

melted butter, for greasing
4 all-purpose potatoes, such as desiree or
    pontiac (about 800 g/1 lb 12 oz)
4 prosciutto slices
1/4 cup chopped mint leaves
150 g (5 oz/1 1/4 cups) fresh sheep's milk
    cheese (or goat's)
250 g (9 oz/1 cup) sour cream
2 eggs, lightly beaten
125 ml (4 fl oz/1/2 cup) pouring
    (whipping) cream
sea salt

Preheat the oven to 180°C (350°F/Gas 4). Brush four 250 ml
(9 fl oz/1 cup) ramekins with melted butter.

For each pie, peel and thinly slice one potato and pat dry
on paper towel. Line the base of a prepared ramekin with half
a slice of prosciutto then arrange half the potato slices over, in
an even layer. Place the remaining halved prosciutto slice over,
then scatter over a quarter of the cheese and a quarter of the
mint leaves. Cover with the remaining potato slices and press
down firmly on the mixture; the ramekin should be filled to the
top. Repeat the process to fill the remaining ramekins.

Combine the sour cream, eggs and cream in a bowl,
season to taste with sea salt and freshly ground black pepper,
then whisk until smooth. Divide the sour cream mixture
among the ramekins, pouring it over slowly to allow it to seep
between the layers.

Place the ramekins on a baking tray, then bake for
50–60 minutes, or until the mixture is bubbling and the potato
is soft when tested with a skewer. Stand for 5 minutes, then
loosen the edges of each pie with a knife. Turn the pies out
onto serving plates and serve immediately.

SMOKED FISH PÂTÉ WITH BRUSCHETTA
MAKES 500 ML (17 FL OZ/2 CUPS); SERVES 6

2 x 400 g (14 oz) hot-smoked rainbow
    trout fillets, skinned and bones removed
2–3 tablespoons lemon juice
125 g (4 oz/½ cup) cream cheese,
    softened
200 g (7 oz) butter, melted and cooled
sprigs of herbs, such as dill, fennel or
    flat-leaf (Italian) parsley, to garnish
lemon slices, to garnish

BRUSCHETTA
1 baguette, sliced diagonally into
    24 thin slices
80 ml (2½ fl oz/⅓ cup) olive oil
3 garlic cloves

Roughly flake the flesh of the fish. Process the flesh in a blender or food processor with the lemon juice, cream cheese and melted butter until the mixture is quite smooth. Season to taste with freshly ground black pepper.

Spoon the mixture into a 500 ml (17 fl oz/2 cup) ramekin, cover with plastic wrap and refrigerate overnight, or until the mixture is firm. Keep covered with plastic wrap and refrigerated until ready to serve. Garnish with sprigs of fresh herbs and lemon slices.

For the bruschetta, preheat the oven to 200°C (400°F/Gas 6). Brush both sides of the bread slices lightly with oil, then spread on a baking tray and bake for 10–15 minutes, or until crisp and golden, turning once. Remove from the oven and rub all over one side of each slice with a garlic clove, using a clove for every 8 slices. Serve with the pâté.

CAPSICUM ROLLED WITH GOAT'S CHEESE, BASIL AND CAPERS
SERVES 4–6

4 large red capsicums (peppers)
¼ cup chopped flat-leaf (Italian) parsley
2 tablespoons snipped chives
2 tablespoons small capers, rinsed, drained
  and finely chopped
1 tablespoon balsamic vinegar
150 g (5½ oz/1¼ cup) crumbled soft
  goat's cheese
16 basil leaves
olive oil, to cover
crusty Italian bread, to serve

Preheat the grill (broiler) to high. Place the capsicums on an oven tray under the grill and cook, turning often, for 10–15 minutes, or until the skin is blackened all over. Place the capsicum in a plastic bag and leave to cool, then peel away the skin. Cut the capsicum into 3 cm (1¼ inch) wide strips.

Combine the parsley, chives, capers and balsamic vinegar in a small bowl. Add the goat's cheese and mix well. Season to taste with freshly ground black pepper. Lay the capsicum pieces, skin side down, on a work surface then place a basil leaf on each capsicum piece. Top each with a teaspoon of the goat's cheese mixture. Roll each piece of capsicum up and over the goat's cheese, to form a roll. Secure with a toothpick. Place in an airtight, non-reactive container and cover with olive oil. Refrigerate until required, then bring to room temperature before serving. Serve with crusty Italian bread.

CHEESE AND SPINACH PANCAKES
SERVES 4

250 g (9 oz) cooked English spinach,
   squeezed dry and chopped (about
   600 g/1 lb 5 oz uncooked weight;
   see Note)
125 g (4 oz/^{1}/$_2$ cup) fresh curd cheese
   (goat's, sheep's or cow's milk)
30 g (1 oz/^{1}/$_4$ cup) grated cheddar cheese
a pinch of freshly grated nutmeg
25g (^{3}/$_4$ oz/^{1}/$_4$ cup) grated parmesan
   cheese
^{1}/$_2$ teaspoon paprika
40 g (1^1/$_2$ oz/^{1}/$_2$ cup) fresh breadcrumbs

BATTER
125 g (4 oz/1 cup) plain (all-purpose)
   flour
310 ml (10^3/$_4$ fl oz/1^1/$_4$ cups) milk
1 egg, lightly beaten
butter, for cooking

CHEESE SAUCE
2 tablespoons butter
30 g (1 oz/^{1}/$_4$ cup) plain (all-purpose) flour
435 ml (15^1/$_4$ fl oz/1^3/$_4$ cups) milk
125 g (4 oz/1 cup) grated cheddar cheese

Grease a large ovenproof dish.

Combine the spinach, curd cheese, cheddar, nutmeg and freshly ground pepper, to taste, in a bowl and mix well.

To make the pancake batter, sift the flour and a pinch of salt into a bowl. Add half the milk and the egg. Whisk until smooth, then whisk in the remaining milk.

Heat a teaspoon of butter in a frying pan over medium heat and slowly pour in the batter to create a thin layer in the base of a pan. Cook for 2–3 minutes, or until golden, then turn over and cook the other side. Repeat to make 8 pancakes.

To make the cheese sauce, melt the butter over low heat, add the flour and cook for 1 minute. Stirring constantly over medium heat, gradually add the milk, bringing the mixture back to a simmer between additions and stirring well to prevent lumps from forming. Cook until the mixture thickens, then remove from the heat, season to taste with salt and pepper and stir in the grated cheese.

Preheat the oven to 180°C (350°F/ Gas 4). Divide the spinach filling among the pancakes, roll the pancakes up then place in the prepared dish. Pour the cheese sauce over the pancakes. Combine the parmesan, paprika and breadcrumbs in a bowl then sprinkle over the sauce. Bake for 30 minutes, or until golden brown. Serve immediately.

NOTE You can use 250 g (9 oz) frozen spinach if preferred; allow it to thaw, then squeeze out all the water.

GIANT CONCHIGLIE WITH RICOTTA AND ROCKET

SERVES 8

32 dried giant conchiglie (shell pasta)

FILLING

500 g (1 lb 2 oz/2 cups) ricotta cheese

100 g (3½ oz/1 cup) grated parmesan cheese

150 g (5½ oz/4½ cups) rocket (arugula), finely shredded

1 egg, lightly beaten

185 g (6½ oz) marinated globe artichokes, drained and finely chopped

80 g (2¾ oz/½ cup) sun-dried tomatoes, finely chopped

95 g (3¼ oz) skinned roasted capsicum (pepper), finely chopped

CHEESE SAUCE

60 g (2¼ oz) butter

30 g (1 oz /¼ cup) plain (all purpose) flour

750 ml (26 fl oz/ 3 cups) milk

100 g (3½ oz) gruyère cheese, grated

2 tablespoons chopped basil, plus extra whole leaves, for garnish

600 ml (21 fl oz) tomato passata (puréed tomatoes)

½ teaspoon finely chopped oregano

Cook the pasta in a large saucepan of boiling salted water until *al dente*. Drain well. Arrange the shells on two nonstick baking trays, spacing them well apart so they don't stick together. Cover lightly with plastic wrap.

To make the filling, combine all the ingredients in a large bowl and season well. Spoon the filling into the shells, taking care not to overfill them or the fillings will spill out as the shells cook.

To make the cheese sauce, melt the butter in a small saucepan over low heat. Add the flour and stir for 1 minute. Stirring constantly over medium heat, gradually add the milk, bringing the mixture back to the simmer between additions and stirring well to prevent lumps from forming. Cook until the mixture thickens. Remove from the heat, stir in the gruyère cheese with the chopped basil and season to taste.

Preheat the oven to 180°C (350°F/Gas 4). Spread 250 ml (9 fl oz/1 cup) of the cheese sauce over the base of a 3 litre (104 fl oz/12 cup) capacity ovenproof dish. Arrange the filled conchiglie over the sauce, top with the remaining sauce and bake for 30 minutes, or until the sauce is golden.

Combine the passata and oregano in a saucepan, cover then bring to a simmer and cook for 3–4 minutes, or until heated through. Divide the sauce among warmed serving plates, top with conchiglie and sprinkle with basil leaves. Serve immediately.

NEW YORK CHEESECAKE
SERVES 8–10

PASTRY

60 g (4 oz/$\frac{1}{2}$ cup) self-raising flour

230 g (8 oz/1$\frac{3}{4}$ cups) plain (all-purpose) flour

60 g (2$\frac{1}{4}$ oz/$\frac{1}{4}$ cup) caster (superfine) sugar

1 teaspoon finely grated lemon zest

80 g (2$\frac{3}{4}$ oz) butter

2 eggs, lightly beaten

FILLING

750 g (1 lb 10 oz/3 cups) curd cheese or cream cheese, softened

230 g (8 oz/1 cup) caster (superfine) sugar

60 g (2$\frac{1}{4}$ oz/$\frac{1}{2}$ cup) plain (all-purpose) flour

2 teaspoons grated orange zest

2 teaspoons grated lemon zest

4 eggs

170 ml (5$\frac{1}{2}$ fl oz/$\frac{2}{3}$ cup) pouring (whipping) cream

glacé (candied) citrus slices, to decorate (optional; see Note)

Combine the flours, sugar and lemon zest in a bowl, add the butter then, using your fingertips, rub in until the mixture resembles coarse breadcrumbs. Add the egg and mix well. Gradually add 3–4 tablespoons of cold water, or enough to give a coarse dough, then turn out onto a lightly floured surface and gather into a ball. Wrap in plastic wrap and refrigerate for 20 minutes, or until firm.

Preheat the oven to 210°C (415°F/Gas 6–7). Roll the pastry between two sheets of baking paper until large enough to fit the base and side of a greased 22 cm (8 inch) round springform cake tin. Ease the pastry into the tin and trim the edges. Line the pastry shell with a piece of greaseproof paper and baking beads (use dried beans or rice if you don't have beads). Bake the pastry for 10 minutes. Remove the paper and beads and bake for a further 5 minutes, or until the pastry is light golden. Cool.

Reduce the oven to 150°C (300°F/Gas 2).

To make the filling, using electric beaters, beat the cream cheese, sugar, flour and zests until smooth. Add the eggs, one at a time, beating well after each addition. Stir in the cream, then pour the filling over the pastry and bake for 1 hour 25 minutes, or until almost set. Cool in the oven, then refrigerate until firm. Serve decorated with glacé citrus slices, if desired.

NOTE Glacé (candied) citrus slices are available from delicatessens and speciality food stores.

PLUM AND CARAWAY BISCUITS
MAKES 24

80 g (2¾ oz) butter, softened
60 g (2¼ oz/¼ cup) cream cheese, at
  room temperature, chopped
115 g (4 oz/½ cup) caster (superfine)
  sugar
1 teaspoon natural vanilla extract
2 egg yolks
1½ teaspoons caraway seeds
150 g (5½ oz/1¼ cups) plain
  (all-purpose) flour
plum jam
icing (confectioners') sugar, for dusting

Cream the butter, cream cheese and sugar in a bowl using electric beaters until pale and fluffy. Add the vanilla and 1 egg yolk and beat to combine well. Add the caraway seeds and flour and stir until a dough forms.

Turn the dough out onto a lightly floured work surface, form into a flat rectangle, then cover with plastic wrap and refrigerate for 2 hours, or until firm.

Preheat the oven to 180°C (350°F/Gas 4). Lightly grease two baking trays. Combine the remaining egg yolk with 2 teaspoons water and stir to combine well.

Cut the dough in half, then roll out each half on a lightly floured work surface to form an 18 x 24 cm (7 x 9½ inch) rectangle. Using a lightly floured sharp knife, cut the dough into 6 cm (2½ inch) squares. Place a scant teaspoon of jam diagonally across the centre of each square, then brush all four corners of each square with the egg mixture. Take one corner and fold it into the centre. Take the opposite corner and fold it into the centre, overlapping the first corner slightly, to partially cover the jam and form a neat oblong shape with pointed ends.

Brush the tops of the biscuits with the egg yolk mixture, then place the biscuits, glazed side up, on the baking trays. Bake for 10–12 minutes, or until light golden, swapping the position of the trays halfway through cooking. Cool on the trays for 5 minutes, then transfer to a wire rack to cool completely. Dust with icing sugar before serving.

The biscuits will keep, stored in an airtight container, for up to 1 week.

DATE AND MASCARPONE TART

SERVES 6–8

90 g (3¼ oz/½ cup) rice flour
60 g (2¼ oz/½ cup) plain (all-purpose)
  flour
2 tablespoons icing (confectioners') sugar
25 g (1 oz/¼ cup) desiccated (grated
  dried) coconut
100 g (3½ oz) marzipan, chopped
100 g (3½ oz) unsalted butter, chilled and
  cut into cubes

FILLING
200 g (7 oz) fresh dates, pitted
2 eggs
2 teaspoons custard powder or vanilla
  instant pudding mix
125 g (4½ oz) mascarpone cheese
2 tablespoons caster (superfine) sugar
80 ml (2½ fl oz/⅓ cup) pouring
  (whipping) cream
2 tablespoons flaked almonds

Preheat the oven to 180°C (350°F/ Gas 4). Grease a shallow
10 x 35 cm (4 x 14 inch) tin.

Combine the flours, icing sugar, coconut and marzipan
in a food processor and process for 10 seconds. Add the butter
and, using the pulse button, process for 10–20 seconds, or
until the dough just comes together. Take care not to over-
process. Turn the dough out onto a lightly floured surface and
gather into a ball. Wrap in plastic wrap and refrigerate for
15 minutes.

Cut the dates into quarters lengthways. Roll out
the pastry between two sheets of baking paper until large
enough to line the tin. Ease it into the tin and trim the edges.
Refrigerate for another 5–10 minutes. Line the pastry shell
with a piece of greaseproof paper and baking beads (use dried
beans or rice if you don't have beads). Bake the pastry for
10 minutes, remove the paper and beads and bake for a further
5 minutes, or until light golden. Cool.

Arrange the dates over the pastry base. Whisk together
the eggs, custard powder, mascarpone, sugar and cream until
smooth. Pour over the dates and sprinkle with almonds. Bake
for 25–30 minutes, or until the top is golden and the filling is
just set. Serve warm.

CANNOLI
MAKES 12

FILLING

500 g (1 lb 2 oz/2 cups) ricotta cheese
   (see Note, page 60)
1/4 teaspoon orange flower water
100 g (3 1/2 oz/ 1/2 cup) cedro, diced (see
   Notes)
60 g (2 1/4 oz) bittersweet chocolate,
   coarsely grated or chopped
1 tablespoon grated orange zest
60 g (2 1/4 oz/1/2 cup) icing (confectioners')
   sugar

DOUGH

300 g (10 1/2 oz/2 1/3 cups) plain
   (all-purpose) flour
1 tablespoon caster (superfine) sugar
1/2 teaspoon ground cinnamon
40 g (1 1/2 oz) unsalted butter
60 ml (2 fl oz/1/4 cup) sweet Marsala
vegetable oil, for deep-frying
icing (confectioners') sugar, for dusting

To make the filling, combine all the ingredients in a bowl and
mix to combine well. Cover with plastic wrap and refrigerate
while you prepare the dough.

To make the dough, combine the flour, sugar and
cinnamon in a bowl, rub in the butter and add the Marsala.
Mix until the dough comes together in a loose clump, then
knead on a lightly floured surface for 4–5 minutes, or until
smooth. Wrap in plastic wrap and refrigerate for at least
30 minutes.

Cut the dough in half and roll each piece out on a lightly
floured surface into a thin sheet about 5 mm (1/4 inch) thick.
Cut each dough half into six 9 cm (3 1/2 inch) squares. Place a
metal cannoli tube (see Notes) diagonally across the middle
of each square. Fold the sides over the tube, moistening the
overlap with water, then press together to seal.

Fill a large deep frying pan one-third full of oil and heat
it to 180°C (350°F), or until a cube of bread dropped into
the pan browns in 15 seconds. Drop one or two tubes at a
time into the hot oil. Fry gently until golden brown and crisp.
Remove from the oil. Gently remove the moulds and drain on
crumpled paper towels. When the pastry shells are cool, fill a
piping (icing) bag with ricotta mixture and fill the shells. Dust
with icing sugar and serve.

NOTES: Cedro, also known as citron, is a citrus fruit with a very
thick, knobbly skin that is used to make candied peel. Cannoli tubes
are available at kitchenware shops. You can also use 2 cm (3/4 inch)
diameter wooden dowels cut into 12 cm (4 1/2 inch) lengths.

GINGER AND RICOTTA PANCAKES WITH HONEYCOMB
SERVES 4

150 g (5½ oz/1 cup) wholemeal
    (whole-wheat) flour
2 teaspoons baking powder
2 teaspoons ground ginger
2 tablespoons caster (superfine) sugar
55 g (2 oz/1 cup) flaked coconut, toasted
4 eggs, separated
500 g (1 lb 2 oz/2 cups) ricotta cheese
    (see Note, page 60)
310 ml (10¾ fl oz/1¼ cups) milk
melted butter, for cooking
4 bananas, sliced

200 g (7 oz) honeycomb confectionery,
    broken into large pieces

Preheat the oven to 120ºC (235ºF/Gas ½).

Sift the flour, baking powder, ginger and sugar into a bowl. Stir in the coconut then make a well in the centre. Add the combined egg yolks, 350 g (12 oz) of the ricotta and the milk then stir until smooth.

Using electric beaters, whisk the egg whites until soft peaks form, then fold into the pancake mixture.

Heat a frying pan over a low heat and brush lightly with a little melted butter. Pour 60 ml (¼ cup) of the batter into the pan and swirl gently to create an even pancake. Cook until bubbles form on the surface and the base is golden, then turn over and cook the other side for 1 minute, or until golden. Repeat until all the batter is used. As each pancake is cooked, transfer to a plate, cover with foil and keep warm in the oven while you cook the remaining pancakes.

To serve, stack three pancakes onto each plate, top with a spoonful of ricotta, some banana and shards of honeycomb.

CHEESE AND SULTANA SLICE
MAKES ABOUT 25

BASE

125 g (4½ oz/½ cup) unsalted butter, cut
  into cubes
70 g (2½ oz) icing (confectioners') sugar,
  plus extra for dusting
185 g (6½ oz/1½ cups) plain
  (all-purpose) flour, sifted

TOPPING

250 ml (9 fl oz/1 cup) milk
30 g (1 oz) unsalted butter
150 g (5½ oz/1¼ cups) crumbled soft,
  mild goat's cheese
100 g (3½ oz) cream cheese, chopped
1 teaspoon finely grated lemon zest
60 ml (2 fl oz/¼ cup) lemon juice
2 tablespoons caster (superfine) sugar
30 g (1 oz/¼ cup) cornflour (cornstarch)
60 g (2¼ oz/½ cup) sultanas (golden
  raisins), chopped
3 egg whites
80 g (2¾ oz/⅓ cup) caster (superfine)
  sugar, extra

Preheat the oven to 180°C (350°F/Gas 4). Lightly grease a
20 x 30 cm (8 x 12 inch) rectangular shallow tin with butter
and line the base with baking paper, leaving the paper hanging
over on the two long sides to aid removal later.

To make the base, cream the butter and icing sugar in a
bowl using electric beaters until pale and fluffy. Add the flour
and stir until a dough forms. Using lightly floured hands,
press the dough evenly over the base of the tin. Bake for
15–20 minutes, or until golden and firm to the touch. Cool.

To make the topping, combine the milk, butter, cheeses,
lemon zest, lemon juice and sugar in a large saucepan and stir
over medium-low heat for 5 minutes, or until the butter and
cheeses have melted and the mixture is smooth. Combine
the cornflour and 60 ml (2 fl oz/¼ cup) cold water in a small
bowl and stir until smooth. Add to the cheese mixture, then,
whisking continuously, bring to the boil and cook, stirring, for
3–4 minutes, or until thickened. Remove from the heat and stir
in the sultanas. Set aside.

Whisk the egg whites in a clean, dry bowl until stiff
peaks form. Add the sugar gradually, whisking well after
each addition. Whisk until the mixture is stiff and glossy and
the sugar has dissolved. Carefully fold the egg whites into
the cheese mixture and mix until just combined. Spread the
mixture over the cooled base. Bake for 25–30 minutes, or until
firm to the touch and light golden.

Cool completely in the tin, then carefully lift out and,
using a hot knife, cut into diamonds. Dust with icing sugar
to serve. The cheese slice is best served on the day it is made.

Tiramisu
Serves 6

5 eggs, separated

175 g (6 oz/³/₄ cup) caster (superfine) sugar

300 g (10½ oz/1⅓ cups) mascarpone cheese

250 ml (9 fl oz/1 cup) cold strong coffee

3 tablespoons Kahlúa or other coffee-flavoured liqueur

36 small savoiardi (lady fingers/sponge finger biscuits)

80 g (3 oz/²/₃ cup) finely grated dark chocolate

Using electric beaters, whisk the egg yolks with the sugar in a bowl until the mixture is thick and pale and leaves a ribbon trail when dropped from the beaters. Add the mascarpone and beat until smooth.

Using clean, dry beaters and bowl, whisk the egg whites until soft peaks form, then fold into the mascarpone mixture.

Combine the coffee and liqueur in a shallow dish. Dip enough biscuits to cover the base of a 25 cm (10 inch) square dish into the coffee mixture, dipping each biscuit for 2–3 seconds; the biscuits should be well soaked but not breaking up.

Arrange the biscuits snugly in the base of the dish. Spread half the mascarpone mixture over the biscuits, smoothing the surface. Dip more biscuits into the coffee mixture and use these to neatly cover the mascarpone layer, then top with the remaining mascarpone mixture, smoothing the surface. Cover the dish with plastic wrap and refrigerate for 3 hours or overnight, to allow the flavours to develop. Sprinkle with chocolate, then serve immediately.

PRUNE AND RICOTTA CAKE
SERVES 8

150 g (5 oz/²/₃ cup) pitted prunes,
  chopped
2 tablespoons marsala
500 g (1 lb 2 oz/2 cups) ricotta cheese
250 g (9 oz) caster (superfine) sugar
3 eggs, lightly beaten
125 ml (4 fl oz/¹/₂ cup) pouring
  (whipping) cream
60 g (2¹/₄ oz/¹/₂ cup) cornflour
  (cornstarch), sifted
60 g (2¹/₄ oz/¹/₂ cup) grated chocolate

Preheat the oven to 160°C (315°F/Gas 2–3). Grease a 23 cm (9 inch) round cake tin and line the base with baking paper.

Combine the chopped prunes and marsala in a small saucepan. Bring to the boil, reduce the heat and simmer for 30 seconds, or until the marsala is absorbed. Allow to cool.

Using electric beaters, beat the ricotta and sugar in a mixing bowl for 4 minutes, or until light and creamy. Gradually add the egg, beating well after each addition. Add the cream and beat for 2 minutes. Gently fold in the cornflour, prune mixture and chocolate with a metal spoon.

Spoon the mixture into the prepared tin and bake for 2 hours, or until firm and cooked when tested with a skewer. Leave in the tin for 15–20 minutes before gently turning out onto a wire rack to cool.

STRAWBERRIES WITH BALSAMIC VINEGAR AND MASCARPONE

SERVES 4

750 g (1 lb 10 oz) ripe small strawberries, wiped clean and hulled
60 g (2 oz/¼ cup) caster (superfine) sugar
2 tablespoons good-quality balsamic vinegar
125 g (4 oz/½ cup) mascarpone cheese, to serve

Halve any large strawberries then place the strawberries in a large bowl. Sprinkle with sugar and toss gently to coat, then stand for 30 minutes.

Sprinkle the vinegar over the strawberries, toss to coat, then cover the bowl with plastic wrap and refrigerate for 30 minutes.

Divide the strawberries among four glasses, drizzle with the syrup from the bowl and top each with a spoon of mascarpone cheese, to serve.

blue and surface-ripened cheeses

There are many cheeses that owe their distinctive flavour, aroma and even appearance to moulds and bacteria. Nowadays, these moulds and bacteria are applied and grown under highly controlled conditions; however, in days gone by, randomly occurring wild spores and bacteria were utilised for cheese making, and the results were not as predictable.

The most famous of such cheeses are the blue cheeses, with their familiar blue-green interior marbling, and the surface-ripened cheeses. The latter ripen from the outside in by the action of special bacteria and under very controlled conditions. There are two types: mould-ripened and washed-rind.

MOULD-RIPENED CHEESES

Perhaps the two most famous examples of this type of cheese are the French cheeses camembert and brie, although there are many others: cambozola, from Germany; bucheron, from France; cabichou and montrachet, both goat's cheeses from France; and domiati, the national cheese of Egypt. Surface-ripened cheeses are among the earliest recorded cheeses made in France. Originally their exteriors — which helped to protect the cheese and extend its life, as well as aiding in the ripening process and lending flavour — were dark grey or black. It wasn't until about 50 years ago that researchers found a way to grow consistently pale moulds on the outside of cheese, and that hue is now strongly associated with these cheeses. Generally, these cheeses are not for cooking, as their special, creamy texture and complex, mellow flavours are lost when subjected to heat. They are best appreciated when fully ripe and at room temperature, slathered over good bread and served with simple accompaniments: fresh fruits (especially apples, pears, grapes or dates) or rich dried fruits, such as figs and muscatels. Tomatoes, slices of prosciutto or other good ham, or a simply dressed green salad are also appropriate accompaniments, as is a glass or two of wine, preferably Champagne. Cooking with these cheeses is really only advised as a way to use any left over that you may not have the chance to eat in its prime.

CAMEMBERT
Camembert is credited as having been invented by a Normandy farmer in 1791 and officially named after its village of origin by Napoleon, when he was served it. Cheese, however, was made in Camembert long before this, and perhaps that now-famous farmer simply refined the process of manufacturing this very popular cheese. Camembert's soft, white, downy, edible rind is formed by the moulds *Penicillium candida* and *P. camberti*, which are allowed to flourish and eventually transform the hard, crumbly interior of the young cheese into a silky soft, pale yellow, creamy one. Today, camembert is made in industrial quantities in nations as diverse as the United States, Denmark and Brazil, although in France, genuine camembert is protected by a strict appellation system and some is still made using raw milk; many brands are sold in the traditional small wooden boxes.

Camembert starts life as any other cheese; milk is curdled, whey is drained off, then the curd is carefully measured into moulds. After applying the mould to the surface, the immature cheeses are left to age for about 12 days, then packaged and shipped for sale. They do not fully ripen until some 30 days after packaging; it is important to note the 'use-by' date, as this will signify the optimum time of ripeness for the cheese. You can also tell by touch and appearance if the cheese is fully ripe. It should feel supple and yielding to the touch when pressed gently, and appear slightly plump. Hard edges or a slightly shrunken-looking crust can signify over-ripeness (over-ripe camembert will have a very runny interior and ammoniac aromas). The cheese ripens because the surface mould spreads microscopic root systems down to the centre of the cheese, breaking the paste down very slowly until the texture is uniformly soft. Unripe camembert will have a chalky, hard interior layer, which entirely disappears when the cheese is ripe. A fully ripe camembert will ooze slightly when cut and have a clear, light yellow colour. Camembert should be stored in the refrigerator but served at room temperature; chilling dulls the flavour and hardens the texture.

A standard 250 g (9 oz) camembert has about 52 per cent fat, 135 g (4³/₄ oz) moisture and 63 per cent proteins and lactose. Sometimes the rind might have light red-brown spots, a sign that various other cultures have been employed to make the cheese. This discolouration is quite normal and the cheese will be perfectly fine to eat.

BRIE

In 1814, after heated debates at a Vienna Congress on the subject of the world's best cheese, a contest was arranged to settle the question — and French brie won. Arguably the best-known and best-loved of all of France's 400-odd cheeses, it originated in the area of the same name, to the east of Paris, but is now made and sold the world over. It is a cow's milk cheese, and old; legend has it that brie was in existence during the time of Charlemagne, around AD 774. It takes about 30 litres (8 gallons) of milk to make one 2–3 kg (4¹/₂–6¹/₂ lb) round of brie. It is made in much the same way as camembert; milk is heated to just 37°C (98.5°F), so as not to 'cook' it, then separated into curds and whey. After this the curds are carefully transferred to moulds, left to drain for 18 hours then taken out of the moulds. It is lightly dry-salted, inoculated with essentially the same moulds as camembert then matured in cool cellars for up to four weeks.

Like camembert, brie is fully ripe when it has no chalky layer left in the middle — it should be soft and yielding to the touch at room temperature and ooze a little when cut. Aromas of truffle and mushroom exude from good brie and the flavours are full, rich and buttery. Because camembert is small and loses moisture more quickly, it can age to have a stronger flavour than brie. To truly experience what these two cheeses should taste like, one must sample unpasteurised French specimens. These exhibit characteristics that are due to the *terroir* (pasture, location, climatic conditions) under which each is made, and these subtleties are largely lost in industrial processing using treated milk. In France, when made according to age-old practices and using raw milk, the two cheeses are quite distinct from each other in taste and texture.

Brie was originally made when there was surplus milk, in spring and early autumn, and in various sizes. It was aged on straw beds in moist cellars and named for its village of production, the most famous being brie de Meaux. Unhappily, the majority of brie and camembert made in industrial countries today has been 'stabilised', meaning that the acid development of the curds and the action of the starters used are highly temperature-controlled operations and the processes are done rather quickly. In these cheeses, there is no chalky interior for the mould to work its way through; rather, from very early on in the process, the cheeses have a uniformly soft centre, which will hold for up to about 70 days. They do not develop much extra character or flavour over this time and the characteristic external bloom exerts very little influence over the taste or texture. Often, cream is added to the milk to give the cheese more flavour; these 'double brie' cheeses have about 60 per cent fat. The advantage of industrially made cheeses is that they are fairly consistent in texture and flavour (whereas traditionally made white-rind cheeses can be highly variable) and are easy to handle and store, requiring no special attention as they ripen. The disadvantage is that they lack the complexity and full taste of traditionally made examples.

WASHED-RIND CHEESES

This family of cheeses comprises the infamous 'stinky' cheeses, renowned for their strong, earthy aromas that can permeate an entire room. The smell comes entirely from the rind; usually, the inside paste is sweet, smooth and nutty in flavour and not nearly as pungent as the rind might imply.

It is thought that washed-rind cheeses were first developed in French monasteries in medieval times. Their strong flavour was desired by the monks, particularly during the meatless days of Lent and other fasting periods. The monks discovered that if the rind of the cheese was washed during aging, the texture of the cheese became moist and supple, due to the action of bacteria on the cheese's damp exterior. The bacteria responsible are *Brevibacterium linens*, and this is what gives many washed-rind cheeses their distinctive orange-coloured rind and their flavour. Like white-mould cheeses, they ripen from the outside in; the action of the bacteria on the surface slowly causes the interior to ripen over time and the cheese to develop rich, complex flavours. These cheeses need a high surface-to-volume ratio so that the very interior of the cheese will ripen, hence they are almost always flattish and disc-like in shape (although some, like taleggio, are square). Their rind must be strong enough to support their ripe, slightly bulging interior without cracking. Many different solutions are used to wash these cheeses; brine, beer, wine and brandy are all common. Salt is always included, to discourage the growth of undesirable bacteria, and the cheese is aged in cool, damp conditions. The more the cheese is washed and the more humid the aging conditions, the stronger it will smell and taste.

Washed-rind cheeses are, like many families of cheese, difficult to accurately categorise due to their variety of forms and textures. Some, such as époisses, from Burgundy, or affidélice, from Chablis, are soft and spoonable when ripe; others, such as morbier and the Swiss raclette, are much firmer. These days, like white-mould cheeses, most washed-rind cheeses are stabilised, resulting in minimal surface ripening and far tamer smells and flavours.

PONT L'ÉVÊQUE

Named after the village in Normandy where it supposedly first appeared, this cheese is one of the most popular in France. An old cheese, it dates from at least the 12th century, when it was made by monks; it was once known as 'white meat' as it was eaten instead of meat on fasting days. Pont l'évêque is a cow's milk cheese, with a full, rich, sweet, slightly tangy flavour profile and a thin brownish rind; it is very similar to livarot, another washed-rind cheese from the same area (see below). Pont l'évêque is made both industrially and artisanally and is aged for at least two months prior to sale. The paste, which contains about 45 per cent fat, is a creamy yellow colour with small holes throughout, and oozes slightly when the cheese is fully ripe. The rind, which takes on a more intensely orange-brown colour as the cheese ages, bears fine ridges from the straw mats the cheese is traditionally cured on. Pont l'évêque is made in three sizes, but the shape is always square. Like other washed-rind cheeses, pont l'évêque is best appreciated when served at room temperature, with simple accompaniments.

LIVAROT

Pungent and earthy, with a striking, sticky orange rind, livarot is one of Normandy's oldest cheeses. During the 19th century, livarot was called 'the meat of the poor' due to its nutritional qualities and its importance in the diets of those who could not afford meat. Made from semi-skimmed cow's milk, livarot is aged for up to two months, during which time it is washed in a solution containing annatto, which gives the rind its characteristic colour.

The paste of livarot is extremely creamy, slightly springy and golden yellow, with small holes throughout. It contains about 40 per cent fat; when fully ripe, the taste and aroma of the cheese are full and piquant.

MUNSTER

Munster is another ancient cheese that began life in French monasteries. Dating to about the 7th century, it comes from Alsace, near the German border. It is made from cow's milk, shaped into rounds and traditionally matured in caves for about three months. These days most Munster is made industrially, from pasteurised milk, although some traditional producers in the Alsace region still use raw milk and age-old processes. One such practice is that of leaving the young cheeses on straw mats in caves near the older cheeses, from which they get the flora that infect their rind; in factories, this is done in a more controlled, and far less romantic, manner. Every second day the rind is washed with a solution containing annatto, which colours the rind orange.

Mature munster has a soft, creamy, very rich paste that some liken to melted chocolate on the palette. Its flavour is strong and piquant, and is best complemented by chunks of crusty baguette or slices of walnut bread, washed down with fruity gewürztraminer or pinot noir.

Munster is not to be confused the modern American cheese Muenster, which is bland in the extreme.

TALEGGIO

Taleggio is among the world's oldest soft cheeses, dating from at least the 11th century AD. It is native to an area near Bergamo, in Italy's north, where farmers traditionally aged it in caves. These days, after milk is curdled and the drained curds are shaped into the characteristic large squares, much taleggio is left to cure on wooden shelves under conditions that only mimic the damp, cool interior of caves. The cheese is brushed regularly with a saline solution to inhibit the growth of unwanted bacteria, and after 35–40 days, the cheese is ready. Although strong-smelling on account of the rind, taleggio has a somewhat sweet and aromatic flavour, with a faint mushroomy aftertaste. As the cheese ripens, the paste changes from white to pale yellow and develops a very smooth, creamy texture. It makes an excellent table cheese, served simply at room temperature with fruits such as apple or pear and some good bread. It is also excellent when used in pasta and rice dishes (such as risotto), or in omelettes and vegetable gratins. Like most other cheeses, beware industrially made taleggio, which possesses none of the character or complexity of 'real' taleggio; there are a handful of producers making it in the old way, aging it in caves.

BLUE CHEESE

The term 'blue cheese' is applied to a number of quite different cheeses, most made from cow's milk, some from goat's or sheep's milk. What they all share is that they have been infected by particular types of mould which, once grown throughout the interior of a cheese, produce a striking network of blue-green mould and also affect the flavour and texture of the cheese.

Complex in flavour and with a pronounced aroma and taste, many blue cheeses (notably roquefort, stilton and gorgonzola) are considered among the world's best cheeses. Most blue cheeses were originally produced in caves, where they were left to cure. It is likely the first of these were made by mistake; cool, damp caves harbour mould spores and ancient cheese makers probably did not intend for these to infest their cheeses. Perhaps some brave soul tasted what was considered to be soiled cheese and pronounced it not only edible but delicious, and a whole new family of cheeses was born.

Moulds are aerobic, meaning they require air to proliferate. This is why blue cheeses are usually pierced through in many places, to allow air to circulate and the mould to grow evenly throughout the cheese.

The famous blues are specific to a geographic region, and like many great cheeses, are protected by appellation systems, ensuring only cheese from that particular location can be called stilton or roquefort, for example. The following are a few of the most famous blue cheeses, but there are countless other types available.

ROQUEFORT

Roquefort is produced from the milk of Lacaune sheep in limestone country in the south of France, where it has been made for many centuries. Pliny the Elder refers in AD 79 to roquefort, or at least to a cheese very like it. For roquefort to be so named, it must be aged in the Cambalou caves of Roquefort-sur-Soulzon; as of 2003, there were just seven producers of authentic roquefort cheese left in the region. The wrapping of authentic roquefort bears a drawing of a red Lacaune sheep.

Traditionally, the mould for the cheese was grown on pieces of rye bread left in the caves for up to eight weeks. AOC regulations in France state that the cheese must be infected with the penicillium moulds present in the caves. These days the mould (*Penicillium roqueforti*), which occurs naturally in the soil in the caves, is grown in laboratories. It is added to the ewe's milk after the milk has been heated to between 28 and 34°C (82.5 and 93° F). The milk is then curdled, moulded and drained, pierced to allow for air circulation, then left for ten days before transportation to the caves to mature. Once there, the cheeses are left exposed to the air for two to three weeks for the fungus to take hold, then they are wrapped and aged for up to another ten months before they are ready to eat.

Roquefort has an intense, complex flavour profile which is by turns creamy, nutty, slightly sweet and salty. Its flavour is undeniably strong and some find it an acquired taste. It partners well with fruits (both dried and fresh, such as pears, apples, figs and grapes), sweet wines such as port and sauternes, and nuts, including walnuts and hazelnuts.

The high cost of authentic roquefort makes it most suitable as cheese to enjoy for its own sake, but it is also superlative in cooked dishes. Try it crumbled over salads, incorporated into dressings and sauces, worked into butter to serve over steak, stirred through cream sauces for pasta or into risotto, or scattered through a potato gratin.

BLUE BRIE

Blue brie exhibits the best of both worlds — the rich, supple creaminess of a mould-ripened cheese and the flavoursome tang of blue cheese. Its mild taste often appeals even to those who dislike the typical assertiveness of blue cheeses. Blue brie was developed in about 1950, in the Bresse area of Burgundy, France, supposedly in response to the growing popularity of imported Italian cheeses. Styles of blue brie are made around the world; the very popular cambozola, produced industrially in Bavaria since 1970, is a camembert–gorgonzola hybrid. These are cow's milk cheeses, many of which have cream added for extra richness and as such are designated double- or even triple-cream cheeses. Their sweet mildness and silky, supple textures mean they are best savoured as eating cheeses, rather than ones to cook with. As with other mould-ripened cheeses, ensure blue bries are fully mature before eating, although many of these industrially produced cheeses are stabilised and won't mature much once packaged for selling.

CABRALES

This cheese, from the Asturias region in northern Spain, is considered by many connoisseurs to be one of the finest blue cheeses in the world. It is also very old, having been made for about 1000 years. It is traditionally made in spring and summer from a combination of goat's, cow's and sheep's milk; locals claim the cow's milk gives acidity, the goat's milk piquancy and the rich sheep's milk a buttery aroma and creamy texture. These days cabrales is made year round, and much of it from cow's milk only.

Cabrales has intense, complex flavours and is not for the faint-hearted. Many cabrales cheeses are deliberately infected with a *Penicillium roqueforti*-type mould and aged in humid, cool caves for up to five months. In such caves, *P. cabralensis* is prolific and some producers don't inoculate their cheese with mould, but instead allow the mould to spontaneously penetrate the cheese. The interior of a mature cabrales is liberally streaked with blue-grey mould and the paste is yellowish white, with pungent aromas. Cabrales partners best with full-bodied red wines or medium-sweet sherry.

STILTON

Stilton has been made in the English counties of Nottinghamshire, Leicestershire and Derbyshire, using local milk, since the early eighteenth century. There are just six producers of this cheese and they turn out over a million stiltons a year between them. Confusingly, the cheese was never actually made in Stilton but became associated with that town because its fame spread from there when it was sold in the town's markets.

Stilton is inoculated with *P. roqueforti* bacteria and aged for nine weeks before it is ready to be sold. At this time its flavour is slightly acid and the texture crumbly. Some prefer it this way, while others enjoy it aged another 4–6 weeks, when the texture is buttery and the flavour smoother and more mellow. Each stilton produced is subject to a visual inspection; a plug of the cheese is removed and a grader assesses the quality. If the cheese passes the inspection it is sold as stilton; if not, it is sold simply as 'blue cheese'. Good stilton will have a rough natural rind.

GORGONZOLA
Gorgonzola is believed to date back to the ninth century. As with other blue cheeses aged in mould-infected caves, the characteristic moulding probably first happened by accident. These days the mould is added at the beginning of the cheese-making process and the cheese is pierced all over to create holes in which the mould can quickly flourish.

An Italian cow's milk cheese, made with mainly pasteurised milk, gorgonzola takes its name from a village outside Milan, and today its zone of production is legally defined. It is made by about 80 producers, who mostly age the cheeses in a specially built underground area in the town of Novara. Gorgonzola takes three to six months to mature. When young, it is creamy and mild-tasting, while older gorgonzola, called *piccante* (spicy) or *naturale*, is darker in colour, with a drier, more crumbly texture, a very well-blued interior and a thicker rind. A special, quick-maturing *dolce* (mild) gorgonzola was developed in the 1930s as an alternative to the pungent original. It is good in cooked dishes, especially in pasta sauces, melted over polenta, stirred into risotto or crumbled over salads. Gorgonzola (either *dolce* or *piccante*, depending upon your preference) is also used in stuffings, either for vegetables or meats, and is melted over some types of pizza. *Dolce* gorgonzola has a higher fat content than piccante and does not keep as well. Purists spurn the *dolce* version, claiming its flavour and overall character are inferior to those of *piccante*.

Try to buy gorgonzola cut from a wheel, rather than in pre-packaged pieces, which tend to lose flavour and condition quickly. Small pieces can be too old by the time you buy them and you have no way of knowing what condition they are in when you purchase them.

Piccante gorgonzola is a superb table cheese; serve it simply, with fresh fruit such as apples, figs, pears, muscatels or dried figs, walnuts in the shell or even drizzled honey. It can also be served with vegetables and salads and pairs well with asparagus, fennel, celeriac, endive, rocket and tomato.

Hot pumpkin with taleggio and herbs
SERVES 4–6

1 kg (2 lb 4 oz) butternut pumpkin
  (squash), cut into 4 cm (1 1/2 inch) cubes
125 g (4 1/2 oz) taleggio or other washed-
  rind cheese, finely sliced
1 tablespoon chopped parsley
1 teaspoon chopped oregano
1 teaspoon thyme
1 teaspoon freshly grated nutmeg

Put the pumpkin in a large steamer and cover with a lid. Sit the steamer over a saucepan or wok of simmering water and steam for 15–20 minutes, or until the pumpkin is nearly tender.

Preheat the oven to 200°C (400°F/Gas 6).

Transfer the pumpkin to an ovenproof dish and bake for 30 minutes, or until it is golden brown. Arrange the cheese on top and bake for a further 3–4 minutes, or until the cheese has melted.

Combine the herbs and nutmeg and sprinkle over the melted cheese. Season well with salt and freshly ground black pepper and serve immediately.

Sweet potato and washed-rind cheese gratin with thyme
SERVES 8

2 tablespoons oil
2 leeks, white part only, chopped
2 garlic cloves, crushed
1 tablespoon chopped fresh thyme
185 ml (6 fl oz/3/4 cup) pouring
  (whipping) cream
185 ml (6 fl oz/3/4 cup) chicken stock
500 g (1 lb 2 oz) all-purpose potatoes,
  such as desiree, peeled and thinly sliced
500 g (1 lb 2 oz) orange sweet potato,
  peeled and thinly sliced
200 g (7 oz) washed-rind cheese, chopped

Preheat the oven to 210°C (415°F/Gas 6–7).

Heat the oil in a frying pan, then add the leeks, garlic and thyme and stir over low heat for 4 minutes, or until the leeks are soft. Add the cream and stock, bring to a boil then remove from the heat.

Grease a shallow 2 litre (70 fl oz/8 cup) capacity ovenproof dish or roasting tin. Place the potato and sweet potato in alternating layers in the dish then pour the cream mixture over, spreading the leeks with a spoon.

Bake for 35 minutes, then scatter the cheese over the top. Bake for another 15 minutes, or until the mixture is golden brown, the potato is just tender and the cheese is bubbling. Serve immediately.

ROAST VEGETABLES WITH POACHED EGG AND CAMEMBERT
SERVES 4–6

12 baby onions

2 bunches asparagus, trimmed and cut into 4 cm (1½ inch) pieces

4 zucchini (courgettes), trimmed and thickly sliced

2 eggplants (aubergines), cut into 2.5 cm (1 inch) pieces

8 garlic cloves

80 ml (2½ fl oz/⅓ cup) olive oil

2 tablespoons lemon juice

4 eggs

250 g (9 oz) camembert cheese, cut into 2.5 cm (1 inch) pieces

Preheat the oven to 200°C (400°F/Gas 6). Peel the onions and trim the root ends, taking care not to cut too much off the root end or the onions will fall apart.

Combine the onions, asparagus, zucchini, eggplant and garlic in a roasting dish, drizzle with oil and toss to coat well. Season to taste with salt and freshly ground black pepper then roast the vegetables for 20 minutes. Drizzle with lemon juice and roast for another 10 minutes.

Bring a large frying pan full of water to the boil, then reduce the heat to a gentle simmer. Crack an egg into a saucer then gently slip the egg into the water. Repeat with the remaining eggs, taking care they don't touch each other, reduce the heat to very low and cook each egg for 3 minutes.

Divide the vegetables among four small ovenproof dishes then arrange the camembert over the top. Bake for 2–3 minutes, or just until the cheese begins to melt. Top each dish with a poached egg, season to taste with freshly ground black pepper and serve immediately.

Blue cheese gnocchi
SERVES 4

500 g (1 lb 2 oz) floury potatoes,
  quartered
150 g (5½ oz/1¼ cups) plain
  (all-purpose) flour

SAUCE
300 ml (10½ fl oz) pouring (whipping)
  cream
125g (4½ oz) gorgonzola cheese, roughly
  chopped
2 tablespoons snipped chives

Cook the potatoes in boiling salted water for 15–20 minutes, or until tender. Drain well, then push through a potato ricer or mash until smooth. Transfer to a bowl, sprinkle with flour and knead with your hands to combine well. Continue kneading until the mixture is firm, smooth and no longer sticky.

Divide the dough into three equal portions, then roll each portion on a lightly floured board into a sausage shape about 2 cm (¾ inch) thick. Cut each into 2.5 cm (1 inch) pieces then, using lightly floured hands, press each gnocchi against the tines of a fork to slightly flatten and mark them (the indentation helps the sauce stick to the cooked gnocchi).

Bring the cream to the boil in a small saucepan. Boil rapidly, stirring constantly, for about 5 minutes, or until the cream has reduced by one-third. Remove from the heat, stir in the cheese and season with salt and freshly ground black pepper to taste.

Bring a large saucepan of salted water to the boil, then add the gnocchi, in batches, and cook for 2–3 minutes, or until the gnocchi rise to the surface. Using a slotted spoon, remove the gnocchi, drain well and transfer to a warm serving dish.

Pour the sauce over the gnocchi, scatter with the chives and serve immediately.

ROQUEFORT SOUFFLÉ
SERVES 4

15 g (¹/₂ oz) butter, melted, for brushing,
  plus 30 g (1 oz) extra
30 g (1 oz/¹/₄ cup) plain (all-purpose) flour
250 ml (9 fl oz/1 cup) milk
125 g (4¹/₂ oz) roquefort cheese
4 egg yolks
a pinch of grated nutmeg, or to taste
5 egg whites

Preheat the oven to 200°C (400°F/Gas 6). Cut a wide strip of greaseproof paper long enough to fold around a 1.25 litre (44 fl oz/5 cup) soufflé dish, then fold in half and tie around the dish so it extends about 5 cm (2 inches) above the top. Brush the inside of the dish and the collar with the melted butter and place the dish on a baking tray.

Melt the remaining butter in a heavy-based saucepan and stir in the flour to form a paste. Stirring constantly over low heat, gradually add the milk, bringing the mixture back to the simmer between additions and stirring well to prevent lumps forming. Simmer, stirring, for 3 minutes, then remove from the heat.

Stir the cheese into the sauce until it melts; do not worry if the cheese separates slightly. Beat in the yolks, one at a time, beating well after each addition. Season to taste with nutmeg, salt and freshly ground black pepper and pour into a large mixing bowl.

Using electric beaters, whisk the egg whites in a clean, dry bowl until soft peaks form. Spoon a quarter of the egg white onto the soufflé mixture and quickly but lightly fold it in, to loosen the mixture. Gently fold in the remaining egg white, then pour the mixture into the soufflé dish.

Bake for 20–25 minutes, or until the soufflé is well risen and wobbles slightly when tapped. Test with a skewer through a crack in the side of the soufflé — the skewer should come out clean or slightly moist. If the skewer is slightly moist, by the time the soufflé arrives at the table it will be cooked in the centre. Serve immediately.

BEETROOT AND BLUE CHEESE SALAD
SERVES 4

1 tablespoon olive oil
50 g (1³/₄ oz/¹/₂ cup) pecans
1.3 kg (3 lb) small beetroot (beets),
  washed, trimmed and halved
250 g (9 oz) baby green beans, trimmed
120 g (4¹/₄ oz/4 cups) watercress, trimmed
2 tablespoons walnut oil
1 teaspoon honey
2 teaspoons finely grated orange zest
1 tablespoon cider vinegar
50 g (1³/₄ oz) firm blue cheese, such as
  stilton, crumbled

Heat the oil in a frying pan over medium-high heat then add the pecans. Cook, stirring often, for 3 minutes, or until lightly toasted, then sprinkle with salt and freshly ground black pepper. Remove from the heat and pour into a bowl lined with paper towels. Drain.

Line a large steamer with baking paper, punch holes in the paper, place the beetroot in the steamer and cover with a lid. Set the steamer over a saucepan of boiling water and cook the beetroot for 30–35 minutes, or until tender when pierced with a knife. Remove from the steamer and cool, reserving the water in the saucepan.

Remove the baking paper from the steamer, add the beans to the steamer, cover and cook for 5–7 minutes, or until just tender. Remove the beans and refresh under cold water.

Peel the beetroot, trim off any excess stem and cut into wedges or chunks. Combine the pecans, beans and watercress in a large bowl. Whisk together the walnut oil, honey, orange zest and vinegar in a bowl, then pour over the salad. Add the beetroot and stir gently to just combine. Season to taste, then transfer to a serving platter and sprinkle the blue cheese over to serve.

Focaccia with gorgonzola and pine nuts
Makes 1

STARTER
100 ml (3^1/2 fl oz) milk, warmed
1 teaspoon honey
^{1}/2 teaspoon dried yeast
75 g (2^1/2 oz) plain (all-purpose) flour

DOUGH
^{1}/2 teaspoon dried yeast
250 g (9 oz/2 cups) plain (all-purpose)
   flour, approximately

150 g (5^1/2 oz) gorgonzola or other
   creamy blue cheese, mashed with a fork
1–3 tablespoons mascarpone cheese
10 sage leaves, roughly chopped
2 tablespoons pine nuts

To make the focaccia, make the starter the day before. Combine the milk and honey in a large bowl with 1^1/2 tablespoons warm water. Sprinkle the yeast over, then leave in a draught-free place for 10 minutes, or until foamy. Add the flour and whisk to form a smooth, thick paste. Cover with plastic wrap and stand at room temperature overnight.

To make the dough, sprinkle the yeast over the starter. Using your fingertips, squeeze the starter to break it up, then gradually add 250 ml (9 fl oz/1 cup) water, stirring to combine it with the starter. Using your hands, mix in 1 teaspoon salt and the flour until a soft dough forms.

Turn the dough out onto a lightly floured work surface and knead for 10 minutes, or until smooth and elastic. Place the dough in a lightly oiled bowl, turning to coat with oil, and cover with a damp tea towel (dish towel). Leave in a draught-free place for 1–1^1/2 hours, or until doubled in size. Gently deflate the dough, turn out onto a lightly floured surface and knead for 1–2 minutes, or until smooth. Transfer to a lightly oiled bowl, turning to coat, then cover the bowl with a damp tea towel (dish towel) and stand the dough in a draught-free place for about 40 minutes, or until doubled in size.

Meanwhile, preheat the oven to 220°C (425°F/Gas 7). Lightly grease a baking tray.

Combine the gorgonzola with enough mascarpone to form a soft, spreadable paste. When the dough has risen a second time, roll the dough out on a lightly floured surface until about 2 cm (^{3}/4 inch) thick, then transfer to the prepared baking tray. Spread the gorgonzola mixture over, scatter over the sage and pine nuts and bake for about 20 minutes, or until golden. Serve warm or cold. Focaccia is best served on the day of making.

Mixed salad with warm brie dressing

Serves 4

half a sourdough baguette
185 ml (6 fl oz/³/4 cup) extra virgin
  olive oil
6 rindless bacon slices (about 425 g/15 oz)
2 garlic cloves
2 baby cos (romaine) lettuce, tough outer
  leaves removed
90 g (3¹/4 oz/2 cups) baby English spinach
  leaves
75 g (2¹/2 oz/³/4 cup) pecans, toasted and
  coarsely chopped
2 French shallots, finely chopped
1 tablespoon dijon mustard
80 ml (2¹/2 fl oz/¹/3 cup) sherry vinegar
300 g (10¹/2 oz) ripe brie cheese, rind
  removed, chopped

Preheat the oven to 180°C (350°F/Gas4).

Thinly slice the baguette on the diagonal then, using
2 tablespoons of the oil, brush the slices all over with oil. Place
the bread on a baking tray and bake for 20 minutes, or until
golden. Place the bacon on a baking tray and bake for
5–6 minutes, or until crisp. Rub the bread all over with a cut
clove of garlic. Break the bacon into pieces then cool. Break
the toast into large pieces if the slices are big.

Rinse the lettuce leaves well, then dry thoroughly.
Combine the leaves in a large bowl with the spinach, bacon,
toast and pecans and toss to combine well.

Pour the remaining olive oil into a frying pan and heat
gently, then add the shallots and cook over low heat for
4 minutes, or until softened. Crush the remaining garlic clove,
add to the pan and cook, stirring, for 2–3 minutes, or until
softened slightly. Add the mustard and vinegar to the pan and
whisk to combine well, then add the brie and stir until the
cheese has just melted. Pour the mixture over the salad, toss
gently, then serve immediately.

GORGONZOLA AND TOASTED WALNUTS ON LINGUINE
SERVES 6

75 g (2½ oz/¾ cup) walnut halves
500 g (1 lb 2 oz) dried linguine
70 g (2½ oz) butter, chopped
150 g (5½ oz) gorgonzola cheese,
  crumbled
2 tablespoons pouring (whipping) cream
155g (5½ oz/1 cup) shelled fresh peas
  (approximately 450 g/1 lb in the pod)

Preheat the oven to 180°C (350°F/ Gas 4). Place the walnuts on a baking tray in a single layer and bake for about 5 minutes, or until lightly toasted. Cool.

Cook the linguine in a large saucepan of rapidly boiling salted water until *al dente*. Drain, then return to the saucepan.

Meanwhile, melt the butter in a small saucepan over low heat and add the gorgonzola, cream and peas. Stir gently for 5 minutes, or until the sauce has thickened. Season to taste. Add the sauce and the walnuts to the pasta and toss until combined well. Serve immediately, sprinkled with freshly ground black pepper.

Red leaf salad with blue cheese dressing
SERVES 4

DRESSING

125 g (4½ oz/½ cup) whole-egg
  mayonnaise
60 ml (2 fl oz/¼ cup) thick (double/
  heavy) cream
1 teaspoon white wine vinegar
1 tablespoon finely snipped chives
50 g (1¾ oz) strong blue cheese, such as
  gorgonzola piccante or roquefort
freshly ground white pepper, to taste

150 g (5½ oz) mixed red salad leaves
  (such as red witlof, radicchio and coral
  red lettuce)
1 small bulb fennel (about 100g/
  3⅓ oz), trimmed and finely sliced
1 small red onion, peeled and finely sliced

For the dressing, combine the mayonnaise, cream, vinegar, and chives in a small bowl. Crumble the blue cheese into the mayonnaise mixture and stir to just combine. Season to taste with salt and freshly ground white pepper.

Wash and dry the salad leaves, then tear the larger leaves in half or thirds. Combine the leaves in a large bowl. Add the fennel and the onion to the leaves and toss to combine. Drizzle the dressing over, toss gently, then serve immediately.

BLUE BRIE AND PEAR TOASTS
MAKES 20

1 baguette, cut into 20 slices
100 g (3½ oz) quince paste
2 corella pears, cored and quartered
150 g (5½ oz) blue brie, thinly sliced
extra virgin olive oil, to serve

Heat the oven grill (broiler) to high. Lay the bread slices on the grill tray and toast both sides under the grill until light golden. Cool, then spread with the quince paste.

Cut each pear quarter lengthways into 5 thin slices and arrange 2 pear slices on each slice of toast. Divide the slices of cheese among the toasts, arranging them neatly over the pears.

Cook the toasts under the hot grill for about 1 minute, or until the cheese is warmed through; it will not melt completely. Drizzle with extra virgin olive oil, season with freshly ground black pepper and serve immediately.

CREAMY BLUE CHEESE DIP WITH PEARS
MAKES 200 G (7 OZ)

150 g (5 oz) creamy blue cheese
20 ml (7 fl oz) thick (double/heavy) cream
3 tablespoons Greek-style yoghurt
2 tablespoons finely snipped chives
4 ripe pears, cored and cut into wedges

Mash the blue cheese with a fork to soften it slightly. Add the cream and yoghurt and season with freshly ground black pepper, mixing until smooth; do not overmix, or it will become grainy and curdled. Spoon into a serving bowl, cover and refrigerate for 20 minutes, or until firm.

Scatter the chives over the dip. Serve with the pear wedges.

Stilton soup
SERVES 4–6

30 g (1 oz) butter
2 leeks, white part only, chopped
1 kg (2 lb 4 oz) potatoes, chopped into
  chunks
1.25 litres (44 fl oz/5 cups) chicken stock
125 ml (4 fl oz/½ cup) pouring
  (whipping) cream
100 g (3½ oz) stilton cheese
thyme sprigs, to garnish

Melt the butter in a large saucepan, add the leek and cook, stirring often, over medium heat for 6–7 minutes, or until softened. Add the potato and chicken stock, bring to the boil and simmer, covered, for 15 minutes, or until the potato is tender.

Transfer the mixture to a blender or food processor and blend or process until smooth.

Return the mixture to the saucepan, add the cream and cheese, then stir over low heat until the cheese has melted; do not allow the mixture to boil. Divide the soup among warmed bowls, garnish with sprigs of thyme and serve immediately.

Figs with gorgonzola and prosciutto
SERVES 4 AS AN ENTRÉE OR AS PART OF AN ANTIPASTO SELECTION

8 large black figs
80 g (2¾ oz) gorgonzola cheese, crumbled
8 thin prosciutto slices (about 120 g/
  4¼ oz)
2 tablespoons extra virgin olive oil
2 tablespoons chopped pistachios

Line a steamer with baking paper and punch holes in the paper. Cut the figs into quarters almost to the base, taking care not to cut all the way through, and gently open up slightly. Divide the cheese among the figs, carefully stuffing it inside the figs. Place the figs in the steamer, cover, and cook over boiling water for 6–8 minutes, or until the figs are tender and cheese is warmed through.

Remove the figs from the steamer and carefully wrap a slice of prosciutto around each one, folding the prosciutto over to fit if necessary. Drizzle with olive oil, season to taste with freshly ground black pepper and sprinkle with the pistachios. Serve at once as an entrée or part of an antipasto selection.

ASPARAGUS AND PONT L'ÉVÊQUE PIE
SERVES 6

800 g (1 lb 12 oz) asparagus
20 g (³⁄4 oz) butter
1 large sheet ready-rolled shortcrust
   pastry
175 g (6 oz) pont l'évêque, livarot or
   other washed-rind cheese, sliced
¹⁄2 teaspoon chopped thyme
80 ml (2¹⁄2 fl oz/¹⁄3 cup) pouring
   (whipping) cream
2 eggs
a pinch of ground nutmeg

Trim the base of the asparagus spears and cut any thick spears in half lengthways. Heat the butter in a large frying pan over medium heat and add the asparagus. Add a tablespoon of water and season with salt and freshly ground black pepper. Cook, stirring, for 3 minutes, or until the asparagus is just tender.

Preheat the oven to 200°C (400°F/Gas 6). Grease a 21 cm (8¹⁄2 inch) fluted, loose-based flan (tart) tin.

Roll the pastry out to a 2 mm (¹⁄8 inch) thick circle with a diameter of about 30 cm (12 inches). Line the flan tin and trim the pastry using kitchen scissors, leaving about 8 cm (3 inches) excess around the edge of the tin. Arrange half the asparagus in one direction across the bottom of the dish. Cover with the cheese, then top with the remaining asparagus, laying it in the opposite direction to the previous layer of asparagus. Scatter the thyme on top.

Combine the cream, 1 egg and nutmeg and season. Pour over the asparagus. Fold the pastry over the filling, forming loose pleats to fit. Lightly beat the remaining egg and use it to brush the pastry. Bake in the oven for 35 minutes, or until the pastry is golden.

STUFFED CHICKEN BREAST WITH TOMATO, WASHED-RIND CHEESE AND ASPARAGUS
SERVES 4

4 large boneless, skinless chicken breasts
100 g (3½ oz/²/₃ cup) semi-dried (sun-blushed) tomatoes
100 g (3½ oz) washed-rind cheese, sliced
200 g (7 oz) asparagus spears, trimmed, halved and blanched
50 g (1¾ oz) butter
375 ml (13 fl oz/1½ cups) chicken stock
2 zucchini (courgettes), cut into 5 cm (2 inch) batons
250 ml (9 fl oz/1 cup) pouring (whipping) cream
8 spring onions (scallions), thinly sliced

Gently pound each chicken breast between two sheets of plastic wrap with a meat mallet until 1 cm (½ inch) thick. Divide the tomato, washed-rind cheese and 155 g (5 ½ oz) of the asparagus pieces among the breasts, placing them in the centre of the chicken. Roll up tightly lengthways, securing along the seam with toothpicks.

Heat the butter in a large frying pan over medium heat. Add the chicken, then brown on all sides. Pour in the stock, then reduce the heat to low. Cook, covered, for 10 minutes, or until the chicken is cooked through. Remove the chicken and keep warm in a low oven.

Meanwhile, bring a saucepan of lightly salted water to the boil. Add the zucchini and any remaining asparagus and cook for 2 minutes, or until just tender. Remove from the pan. Whisk the cream into the frying pan. Add the spring onion and simmer over medium-low heat for 4 minutes, or until reduced and thickened. To serve, cut the chicken rolls in half on the diagonal and place on serving plates. Spoon the sauce over and serve with the zucchini, remaining asparagus and spring onions.

POTATO AND GORGONZOLA PIZZA
SERVES 4

BASE
2 teaspoons dried yeast
$^1/_2$ teaspoon sugar
185 g (6$^1/_2$ oz/1$^1/_2$ cups) white strong (bread) flour
150 g (5$^1/_2$ oz/1 cup) wholemeal (wholewheat) plain (all-purpose) flour
1 tablespoon olive oil

TOPPING
1 large red capsicum (pepper)
1 large potato
1 large onion, sliced
125 g (4$^1/_2$ oz) gorgonzola cheese, crumbled
35 g (1$^1/_4$ oz/$^1/_4$ cup) capers, drained and rinsed
1 tablespoon dried oregano
1 teaspoon olive oil

Mix the yeast, sugar, a pinch of salt and 250 ml (9 fl oz/1 cup) warm water in a bowl. Leave in a warm, draught-free place for 10 minutes.

Sift both flours into a bowl. Make a well in the centre, add the yeast mixture and, using your hands, mix until a firm dough forms. Knead the dough on a lightly floured surface for 5 minutes, or until smooth and elastic. Place in a lightly oiled bowl, cover with plastic wrap or a damp tea towel (dish towel) and leave in a warm, draught-free place for 1–1$^1/_2$ hours, or until doubled in size.

Preheat the oven to 200°C (400°F/Gas 6). Brush a 30 cm (12 inch) pizza tray with oil.

Gently deflate the dough, knead for 2 minutes then roll out to a 35 cm (14 inch) round. Put the dough on the tray and press the edge over to form a rim.

To make the topping, cut the red capsicum lengthways into large pieces and remove the membrane and seeds. Place, skin side up, under a hot grill (broiler) until blackened. Cool in a plastic bag, then peel away the skin and cut the flesh into narrow strips.

Cut the potato into paper-thin slices and arrange over the base of the pizza. Top with the capsicum, onion and half the cheese. Sprinkle with the capers, oregano and 1 teaspoon freshly ground black pepper and drizzle with oil. Brush the edge of the crust with oil and bake for 20 minutes. Add the remaining cheese and bake for 15–20 minutes, or until the crust has browned. Serve immediately, cut into wedges.

LEEK, TALEGGIO AND APPLE RISOTTO
SERVES 6

1.25 litres (44 fl oz/5 cups) chicken or
  vegetable stock
2 tablespoons extra virgin olive oil
2 tablespoons unsalted butter
2 leeks, trimmed and cut into
  5 mm (¼ inch) thick rounds
400 g (14 oz/1¾ cups) arborio rice
2 granny smith apples, halved lengthways,
  cored and thinly sliced
250 ml (9 fl oz/1 cup) dry white wine
200 g (7 oz) taleggio cheese, chopped
3–4 sage leaves, finely chopped, plus
  extra, to garnish

Heat the stock in a saucepan over medium heat and keep to a gentle simmer.

Heat the oil and butter in a large saucepan, add the leek and cook over medium-low heat for 4–5 minutes, or until softened. Add the rice and apple and cook, stirring, for 2–3 minutes, or until the rice is well coated and heated through. Add the wine and cook, stirring, until the wine is absorbed.

Add the simmering stock to the rice mixture, 1 cup at a time, stirring constantly until the stock is absorbed before adding any more, until the rice is creamy and tender.

Remove from the heat, stir in the taleggio and chopped sage and season to taste with sea salt and freshly ground black pepper. Divide among warmed bowls, decorate with sage leaves and serve immediately.

CAMEMBERT WITH PORT-SOAKED RAISINS
SERVES 4

2 tablespoons raisins
2 tablespoons port
365 g (12¾ oz) whole camembert cheese
unsalted butter, for greasing
walnut bread, toasted, to serve

Combine the raisins and port in a small saucepan, heat until the mixture just comes to the boil, then remove from the heat and cool for 30 minutes.

Preheat the oven to 180°C (350°F/Gas 4).

Cut a circular lid from the top of the camembert, leaving a 2 cm (¾ inch) border around the edge. Carefully remove the lid, and scoop out the cheese paste with a teaspoon, taking care to leave the base and side intact.

Place the raisins in the cavity of the cheese, then spoon in the cheese paste, pressing down firmly so that it fits neatly into the cavity, then replace the lid.

Lightly grease a doubled-over piece of foil, large enough to contain the camembert, then wrap the camembert tightly in the foil, folding the edges to seal. Place on a baking tray and bake for 10 minutes, or until just heated through. Serve immediately with slices of toasted walnut bread.

semi-soft, goat's milk and sheep's milk cheeses

Semi-soft cow's milk cheeses, and goat's and sheep's milk cheeses, cover a huge variety of types, shapes and textures. Industrial production of cheeses traditionally made with goat's or sheep's milk, often replaces those made with cow's milk, which is cheaper. It is worth seeking out artisanal producers to taste the real thing.

SEMI-SOFT COW'S MILK CHEESES

This family includes the French morbier and laguiole, the Spanish cantabria, the Italian fontina, and the German tilsit, among many others. Three of the most popular types are havarti, monterey jack and bel paese.

HAVARTI

Havarti, one of the many Danish cheeses, was named after the farmhouse where it was first made in the 19th century. It is a semi-soft, rindless, interior-ripened cheese, meaning that ripening starts from the very interior of the cheese, working outwards. Once fully ripened, such cheeses are sealed in some way to prevent any action of oxygen on the exterior which might cause over-ripening — havarti is often sold in a foil wrapping. Havarti has a pale yellow, shiny interior that is supple in texture and is full of irregular, small holes, giving it a lacy appearance. Very mild tasting with a slight acidic edge when young, the cheese acquires a stronger, saltier flavour when older; it is sometimes flavoured with caraway seeds. Cream havarti is a variety with cream added to give an extremely buttery and rich texture; this cheese is very mild in flavour and it is best served as a table cheese, with fruits and wine. Conventional havarti can be used in cooking (it melts readily), but it should not be paired with strong-tasting ingredients as these will overpower its flavour. It is also a good choice for the table, where it will be appreciated by those who don't enjoy strong-flavoured cheeses.

MONTEREY JACK
Monterey jack has been described as the most significant cheese to ever be developed in America. It was supposedly named after both the Californian town Monterey and a nineteenth-century Californian farmer, David Jacks, who started making the cheese on his properties near the town and shipped it with the name 'Monterey Jack's'. The cheese was a huge success and eventually the 's' was dropped and the name stuck. Jack cheese is thought to be based on cheeses made in church missions in California 200 years ago and has been produced commercially for some 100 years — more than a third of all California cheese makers make monterey jack. A cooked-curd cheese, monterey jack is semi-soft when young (aged for about a month) but firms and sharpens in taste with further aging (about six months), becoming crumbly and a little pungent, like cheddar. 'Dry' jack is firmer still, and was developed during the Second World War in response to wartime shortages of parmesan. Monterey jack, when young, has high moisture content, is soft-textured and tastes subtly butter-like and a little sweet. It melts easily and is often employed in dishes such as enchiladas, nachos and tacos, where a mild, melting cheese is required. When young it is like muenster, another popular American semi-soft cheese that has a mild, and some say bland, flavour.

MOZZARELLA (INCLUDING BOCCONCINI)
These cheeses could also be discussed in the Soft and Fresh Cheeses chapter, as they are highly perishable and designed to be eaten quickly; in Italy, fresh mozzarella is considered to be at its best when eaten a day after making. With their firmer, slightly rubbery textures, they can also be described as 'semi-soft'. Their unique method of manufacture sets them apart from other fresh cheeses.

Like that of many great cheeses, the story of mozzarella is shrouded in legend. One story suggests that it was made by accident when a batch of curds fell into a pot of hot water in a cheese factory near Naples, in the south of Italy. Mozzarella was indeed first made around Naples, from the milk of water buffalo. A cow's milk version has also long been made and this is known as *fior de latte*, or 'flower of the milk'. Because it is so highly perishable, it was not much known outside of the south of Italy until relatively recent times, and then only thanks to refrigerated transport. Now it is made all over

Italy and in many other countries too, although the very best mozzarella is considered to be that produced artisanally, by tiny producers, in an area just south of Naples. Pure buffalo-milk mozzarella was once the food of the poor but is now prized the world over; this is the mozzarella the Italians consider to be 'real'. A little of this cheese is exported, so one can taste it without going to Italy, although it can be hard to find and is expensive. With its deliciously soft texture and subtle, milky flavour, mozzarella is a cheese like no other. There are two main types — a low-moisture one (moisture content less than 50 per cent) and a higher moisture one (moisture content more than 52 per cent). The former is a fairly cheap, mass-produced cheese that was developed in America for the pizza industry. This has a longer shelf life than 'true', high-moisture mozzarella and is an inferior cheese in every way.

Mozzarella is initially made in the way customary for most cheeses — pasteurised milk is coagulated, either by the introduction of an acid or by the use of a culture. When the curds reach a pH of 5.2, they are chopped and mixed with hot water, which 'cooks' them. They are then stretched or spun to form long ropes. This process is called *pasta filata* and is unique to a handful of cheeses, provolone and scamorza among them. The curds are thus worked until they are very smooth and elastic, then formed into balls and put into cold water to help them keep their shape as they cool. It takes just eight hours to transform milk into mozzarella. Once formed, the cheese is salted and packaged — sometimes in a vacuum-sealed plastic package, which dramatically increases shelf life, but more often in whey or water, in which it is sold in plastic pots or by weight from a cheese shop or counter. Occasionally the cheese is smoked or flavoured with herbs or pepper. Making good mozzarella involves enormous skill and attention on the part of the cheese maker, which should explain why good mozzarella is so costly. The curds have to be melted at just the correct temperature and the cheese maker must choose just the right time to begin the stretching process. This process causes a thin skin, formed by the outer layer of the stretched cheese, to form around the curds; a cross section of well-constructed mozzarella shows layers of spun or pulled curd, enclosed within a thin skin. Much mozzarella is made by machine, although the very best is still handmade. Good mozzarella should be a bright, white colour, although it can sometimes have a yellowish cast, due to seasonal variations in pasture. It should be delicately milky in both smell and taste — if it tastes at all sour, it is too old. The freshest mozzarella is rather elastic in texture — it softens as it gets older. Mozzarella is most popularly used in salads, especially in the classic combination comprising tomatoes, basil and mozzarella. Served in slices, it is also a welcome addition to an antipasto selection; it pairs extremely well with the salty, assertive tang of ingredients such as salami, dried tomatoes, chargrilled eggplant, olives and anchovies. It is also used on pizzas, where it is cooked just to the point of being warmed through and melting; when melted, it becomes deliciously stringy, but care should be taken not to overheat it or its texture will be destroyed. If using it on pizza or other hot dishes, add it at the last moment, or simply take care not to overdo the cooking time. Mozzarella is excellent in salads, where its bland, sweet, milky flavours and accommodating texture are a counterpoint to ingredients such as juicy tomatoes, salty olives, prosciutto or zingy herbs such as oregano and basil. Bocconcini, which means 'small mouthfuls' in Italian, are small balls of mozzarella usually used in salads or other cold dishes. They are egg-sized (although even smaller bocconcini, sometimes called cherry bocconcini, are also available) and are sold in whey or water, either by weight from cheese shops or packaged into plastic pots.

Bel paese

This semi-soft cheese from the Lombardy region of Italy was invented in the 1920s in a quest by cheese maker Egidio Galabani to make a cheese that would compete with the lush French cheeses becoming so popular at the time. The name 'bel paese' means 'beautiful country' and was inspired by a book of the same name whose author, Stoppani, was a friend of Galabani's.

The cheese itself has elegant and mild flavours (some call it bland) and a soft, buttery-smooth texture; it is grouped with other cheeses of the so-called 'butter cheese' family, of which bella milano, savoie and schonland are other examples. It has a thin, shiny rind and an ivory paste with small holes distributed throughout. A cow's milk cheese with about 50 per cent fat, bel paese is made from pasteurised milk and the manufacturing process is relatively quick. Bel paese is suitable both as a table cheese and for use in cooking; it bears some resemblance to mozzarella when heated, even though the two cheeses are made in very different ways. It can be used in place of mozzarella if that cheese is unavailable. Bel paese melts into ivory smoothness, lending itself to use in risottos, potato and other gratins, in creamy sauces, in ravioli fillings and even in toppings for pizza. It also partners well with crisp, crunchy fruits (such as apples, pears, non-astringent persimmons and grapes) and crackers as an after-dinner course or a snack.

GOAT'S MILK CHEESES

Goat's cheeses vary in texture from soft and spreadable to hard and grateable, and in flavour from mild and sweet to deep and bitingly strong. Goats were one of the world's earliest domesticated animals; about 8000 BC, nomads in the Eastern Mediterranean began keeping goats so they could regularly utilise their meat and milk. Being small and hardy, goats are easy to herd and transport and can withstand life in difficult environments. They also produce more milk for their body weight than any other dairy animal.

The Cyclops Polyphemus in Homer's *Odyssey* is described moulding goat's milk cheese in rush-leaf moulds, and the Romans made goat's cheese, too. It is believed that the Saracen invasion of Spain and France helped spread the practice of farming goats and making cheese from their milk throughout the Mediterranean region. The Saracens got as far as the southern banks of the Loire River, then a major trade and transport route. Today, some of the world's best-loved goat's cheeses — including chabichou, crottin and valençay — come from the area in France south of the Loire.

Goat's milk, and by extension goat's cheese, has a distinctive tang and aroma due to its higher proportions of certain fatty acids. The fat content is slightly higher than that of cow's milk, as are the levels of vitamins B6 and A, potassium, niacin and copper. It contains much less folic acid and vitamin B12 than cow's milk, however, and slightly less lactose, but about the same amount of proteins. The proteins in goat's milk differ slightly from those in cow's milk, allowing goat's curd to be frozen — this is of huge advantage in cheese making, as goat's milk is only seasonally available, in spring and summer.

Goat's cheese is often sold as *chèvre*, which is simply the French word for 'goat'. Goat's cheese is made in all the same categories as cow's milk cheese — fresh and unripened, soft, semi-soft, surface-ripened, semi-hard and hard. Younger goat's cheeses are whiter than aged ones, which turn an ivory colour as they mature. Harder cheeses with less moisture have a stronger flavour and are drier textured, while the freshest ones, designed to be eaten soon after making, are mildly tangy and fresh tasting. Some are sold blended with ingredients such as black pepper, chopped herbs or olives. Soft-ripened goat's cheeses resemble camembert or brie, with a soft, white, edible mouldy crust and, when at their peak, a soft, oozing interior. Both raw and pasteurised goat's milk are used in cheese-making. France, with more than 100 varieties, is home to the widest range of goat's milk cheeses. Aside from traditional European goat's milk cheeses, many boutique-style examples are made in the United States, Canada, Britain and Australia.

SHEEP'S MILK CHEESES

Middle Eastern nomads were the first to milk sheep, and gradually their use as dairy animals spread throughout Western Europe. Sheep's milk contains twice as much fat (9 per cent) as cow's milk. This fat is suspended throughout the milk in tiny globules, which makes it much easier to digest than cow's milk. It takes 4–5 kg (9–11 lb) of ewe's milk, compared to about 10 kg (22 lb) of cow's milk, to make 1 kg (2 lb 4 oz) of cheese.

Due to its much higher casein (milk protein) content, sheep's milk coagulates more quickly than cow's milk, and it produces a firmer curd. Milking sheep produce just 1–2 litres (¼–½ gallon) per day, however, compared with the generous yield from a cow or even a goat, and they produce milk only for two to three months of the year. The milk keeps very well, though; it will hold for several days without spoiling, and is not easily tainted. It is very white (like goat's milk) due to the absence of carotene, and is somewhat thick and viscous in texture. Its strong, distinctive flavour makes it well suited to assertive flavoured, well-salted, aged cheeses, although it also makes delightfully floral-tasting fresh cheeses with high acid levels.

Like the world of cheese itself, that of sheep's milk cheese is vast and complex; for example, in Spain alone there are some 30 different breeds of milking sheep, and the flavour of the milk varies slightly between breeds. Some of the best-known sheep's milk cheeses are feta and haloumi (see below), the various Italian pecorinos, roquefort (see page 151), the Spanish cabrales (see page 152) and manchego, and the Greek kasseri and kefalotyri.

FETA

Although this cheese is most associated with Greece, it is believed that its name comes from the Italian word *fetta*, or 'slice', presumably because that is how feta is served. This is the most consumed cheese in Greece, with an estimated annual consumption of about 10 kg (22 lb) per person. It is also made in many of the countries surrounding Greece, including Bulgaria (which is famed for the quality of its feta), Macedonia, Romania, Serbia, Turkey and Albania. Feta is an ancient food item, widely considered to be one of the world's oldest cheeses; its production goes back at least 5000 years. It is thought that the Cyclops preparing cheese in his cave in Homer's *Odyssey* was making feta, and in more contemporary records (from the Byzantine Empire) there are references to feta cheese.

Technically, feta is a preserved (or 'pickled') fresh cheese, which is why, in ancient times, it could be made and stored without the help of refrigeration. It is made by adding rennet to milk, then hanging the curds to drain and firm. The cheese is then cut into squares or slices and cured in brine — it takes from a week up to several months for feta to age. It is a rindless cheese, with a fat content that can be as low as 30 per cent or as high as 60 per cent. The many styles also vary in flavour, texture and colour depending upon where each is made, from what sort of milk, and the season and method of manufacture. Some feta is sharp and salty, while other fetas are smooth and mild. Feta is traditionally made from a combination of sheep's and goat's milk, aged in wooden barrels and brined so that it will keep; a high percentage of goat's milk makes for a firm, chalkier feta, while more sheep's milk makes feta creamier and smoother in texture. These days the majority of feta comes in large tins, not barrels, as barrel-ripened feta is more time consuming and difficult (and therefore more expensive) to manufacture. Many commercially produced fetas are made using cow's milk, but the flavour is nowhere near as interestingly sharp and tangy as that of feta made from the traditional mixture of milks.

Feta is used widely in cooking, notably in savoury pastries and in egg, vegetable and even seafood dishes from Greece and the Balkans. It is also excellent in salads, or simply nibbled in chunks as an appetiser with pickles or with preserved meats and crusty bread. Feta dries out very quickly once removed from its brine, and exposure to air can make it taste sour. Buy it loose, in its own brine, then store it in brine in the refrigerator at home. You can make brine easily by combining 240 g (8 1/2 oz/3/4 cup) salt with 2 litres (70 fl oz/8 cups) of water; stir to dissolve the salt. Make sure the feta is covered by the brine. It will keep, thus stored, in the refrigerator for some 3 months. Rinse it under cold water to get rid of excessively salty flavours. Feta can also be stored in olive oil to cover, in which case there is no need to refrigerate it, as long as you place it somewhere cool and dark. Feta marinated in olive oil is delicious in salads. If you do purchase feta in plastic packaging, make sure it has some brine around it inside the plastic to protect it from drying out; once opened, use the cheese as soon as possible.

HALOUMI

A cheese from Cyprus, haloumi is traditionally made using a combination of sheep's and goat's milk. Mass-produced haloumi is made using a proportion of added cow's milk, which makes the production cheaper and also alters the flavour. Haloumi has a texture similar to that of mozzarella, although its flavour (mellow but rather salty) is quite different. Also in common with mozzarella, haloumi is a cooked-curd cheese; after it has been formed into curds and then shaped, it is cooked in hot whey or water, which 'sets' the proteins. This means that the cheese has a very high melting point and can be pan-fried or grilled — which it commonly is. One of the most famous dishes involving haloumi is the Greek dish saganaki, wherein it is sliced, lightly floured, then pan-fried until golden and simply served with lemon juice and perhaps with some fresh oregano and black olives scattered over. Haloumi is most conveniently purchased in plastic packs, although some specialty stores sell it by weight, still sitting in its liquid.

GOAT'S CHEESE SCONES
MAKES 10

250 g (9 oz/2 cups) plain (all-purpose) flour
1 tablespoon baking powder
a pinch of cayenne pepper, or to taste
30 g (1 oz) unsalted butter, chilled and cut into cubes
185 ml (6 fl oz/³/4 cup) milk

FILLING
80 g (2³/4 oz/²/3 cup) crumbled goat's cheese
40 g (1¹/2 oz/¹/2 cup) grated pecorino cheese
2¹/2 tablespoons chopped mint

Preheat the oven to 220°C (425°F/Gas 7). Grease a baking tray or line it with baking paper.

Sift the flour, baking powder, cayenne and a pinch of salt into a large mixing bowl. Rub in the butter until the mixture resembles breadcrumbs. Add the milk and, using a flat-bladed knife, mix to form a soft dough. Add a little extra flour if the dough is too sticky. Turn the dough out onto a floured work surface and gently roll out to form a 20 x 25 cm (8 x 10 inch) rectangle. Sprinkle the goat's cheese over, then sprinkle over the pecorino and mint. Starting from the long side, roll the dough into a cylinder. Cut the cylinder into 10 slices, each about 2 cm (³/4 inch) thick. Transfer the slices to a baking tray, spacing them 2 cm (³/4 inch) apart to allow for spreading.

Bake for 10–12 minutes, or until golden and cooked through. Transfer to a wire rack. Serve warm.

HALOUMI, TOMATO AND BASIL SKEWERS
MAKES 22

500 g (1 lb 2 oz) haloumi cheese
5 large handfuls basil
150 g (5¹/2 oz/1¹/4 cups) semi-dried tomatoes
2 tablespoons balsamic vinegar
2 tablespoons extra virgin olive oil
1 teaspoon sea salt

Soak 22 small wooden skewers in cold water for 30 minutes to prevent them from scorching.

Preheat a barbecue hotplate or chargrill pan. Cut the cheese into 1.5 cm (⁵/8 inch) pieces. Thread a basil leaf onto a skewer, followed by a piece of haloumi, a semi-dried tomato, another piece of haloumi and another basil leaf. Repeat to use the remaining ingredients.

Place the skewers on the barbecue hotplate and cook, turning occasionally, until the cheese is golden brown, brushing with the combined vinegar and oil while cooking. Sprinkle with the salt and serve hot or warm.

Goat's cheese scones

Greek lemon, dill and feta bread
MAKES TWO 20 x 10 CM (8 x 4 INCH) LOAVES

375 g (13 oz/3 cups) white bread (strong) flour
125 g (4 1/2 oz/1 cup) semolina
1 tablespoon (12 g/1/4 oz) instant dried yeast
1 teaspoon caster (superfine) sugar
2 tablespoons olive oil
60 g (2 1/4 oz/1 1/2 cups) finely chopped dill
grated zest of 1 lemon
200 g (7 oz/1 1/3 cups) crumbled feta cheese

This bread can be made with an electric mixer or by hand. If making it with a mixer, combine the flour, semolina, yeast, sugar and 1 1/2 teaspoons salt in the bowl of an electric mixer with a dough hook attachment and make a well in the centre. Pour the oil and 250 ml (9 fl oz/1 cup) warm water into the well. With the mixer set to the lowest speed, mix for 3 minutes, or until a dough forms. Increase the speed to medium, add the dill and lemon zest and knead for another 8 minutes, or until the dough is smooth and elastic. Add the feta and knead for 2 minutes, or until the feta is incorporated into the dough.

Alternatively, mix the dough by hand using a wooden spoon, then turn out onto a floured work surface, sprinkle the dill and lemon zest over and knead for 8 minutes, or until the dill and zest are incorporated and the dough is smooth and elastic. Pat the dough into a rectangle approximately 20 x 10 cm (8 x 4 inches) and sprinkle the feta over it. Fold the dough over several times, then knead for 2 minutes, or until the feta is incorporated.

Transfer the dough to a large greased bowl, turning to coat, then cover with plastic wrap and leave in a draught-free place for 1 1/2–2 hours, or until the dough has doubled in size.

Grease two 20 x 10 cm (8 x 4 inch) loaf tins.

Knock back the dough gently, then turn out onto a floured work surface. Divide the dough in half, form each half into a loaf shape then place, seam side down, into the prepared tins. Cover with a damp cloth and leave for 30 minutes, or until doubled in size. Meanwhile, preheat the oven to 200°C (400°F/Gas 6).

Bake the bread for 10 minutes, then reduce the oven to 180°C (350°F/Gas 4) and bake for a further 20 minutes, or until the loaves are golden and sound hollow when tapped on the base. Transfer to a wire rack to cool.

GOAT'S CHEESE FRITTERS WITH ROASTED CAPSICUM SAUCE
MAKES 30

ROASTED CAPSICUM SAUCE
2 red capsicums (peppers)
2 tablespoons olive oil
1 small red onion, finely chopped
1 garlic clove
80 ml (2½ fl oz/⅓ cup) chicken or
  vegetable stock

FRITTERS
425 g (15 oz/1¾ cups) ricotta cheese,
  well drained (see Note, page 60)
400 g (14 oz) firm goat's cheese, crumbled
2 tablespoons snipped chives
30 g (1 oz/¼ cup) plain (all-purpose) flour
2 eggs, lightly beaten
100 g (3½ oz/1 cup) dry breadcrumbs
oil, for deep-frying

Cut the capsicums lengthways into quarters, removing the seeds and membrane. Place, skin side up, under a hot oven grill (broiler) until the skin blackens and blisters. Cool in a plastic bag, peel, then coarsely chop the flesh.

Heat the olive oil in a frying pan over medium heat and cook the onion and garlic for 5 minutes, or until softened. Add the capsicum and stock. Bring to the boil, then remove from the heat, cool slightly and transfer to a food processor. Using the pulse button, process until combined but still a little lumpy. Season to taste and refrigerate until needed.

Combine the ricotta, goat's cheese and chives in a bowl. Add the flour and eggs, then season and mix well.

Put the breadcrumbs in a bowl. Roll a tablespoon of the cheese mixture into a ball with damp hands, flatten slightly and coat in the breadcrumbs. Repeat with the remaining mixture. Refrigerate for 30 minutes.

Fill a deep heavy-based saucepan one-third full of oil and heat to 180°C (350°F), or until a cube of bread browns in 15 seconds. Cook the fritters in batches for 1 minute, or until browned, then drain. Serve warm with the capsicum sauce.

Zucchini and haloumi fritters
MAKES 36; SERVES 6

300 g (10½ oz) zucchini (courgettes)
4 spring onions (scallions), thinly sliced
200 g (7 oz) haloumi cheese, coarsely
  grated
30 g (1 oz/¼ cup) plain (all-purpose) flour
2 eggs
1 tablespoon chopped dill, plus extra
  sprigs, to garnish
60 ml (2 fl oz/¼ cup) vegetable oil
1 lemon, cut into thin wedges
90 g (3¼ oz/⅓ cup) Greek-style yoghurt

Preheat the oven to 120°C (235°F/Gas ½).

Coarsely grate the zucchini then, using your hands, squeeze out as much liquid as possible. Combine the zucchini with the spring onion, haloumi, flour, eggs and dill. Season well with salt and freshly ground black pepper.

Heat the oil in a large heavy-based frying pan. Drop heaped teaspoons of mixture into the pan and cook, in batches, for 2 minutes each side, or until golden and firm. Drain on paper towels, then transfer to a low oven to keep warm while cooking the remaining fritters.

Serve the fritters immediately with yoghurt, a piece of lemon and a small sprig of dill.

NOTE: The fritters are best prepared and cooked as close to the serving time as possible, or the haloumi becomes tough as it cools.

POTATO, FETA AND ROAST GARLIC PASTIES
SERVES 8

PASTRY

400 g (14 oz/3¼ cups) plain (all-purpose)
flour
240 g (8½ oz/1 cup) unsalted butter,
chilled and cut into cubes

FILLING

300 g (10½ oz) all-purpose potatoes such
as pontiac, unpeeled
8 garlic cloves
2 teaspoons rosemary leaves, chopped
2 tablespoons extra virgin olive oil
80 g (2¾ oz/½ cup) crumbled feta cheese
½ teaspoon grated lemon zest
sea salt
1 egg yolk
1 tablespoon milk

For the pastry, sift the flour and a pinch of salt into a large bowl and add the butter. Using your fingertips, rub in the butter until the mixture resembles coarse breadcrumbs. Make a well in the centre and pour in 120 ml (4 fl oz/½ cup) chilled water. Stir with a flat-bladed knife until a coarse dough begins to form. Transfer to a lightly floured work surface and gently knead until the dough comes together. Shape into a flat disc, wrap in plastic wrap and refrigerate for 30 minutes.

Preheat the oven to 180°C (350°F/Gas 4). Lightly grease a baking tray.

Boil the potatoes in their skins for 15 minutes, or until just cooked. Drain the potatoes, cool, then peel and cut into 1 cm (½ inch) pieces.

Put the garlic, rosemary and 1 tablespoon of the oil onto a piece of foil large enough to enclose them, then twist the edges together to make a secure package. Place on the baking tray and roast for 30 minutes. Cool, then squeeze out the garlic from its skin and roughly chop.

Add the garlic to the potato, along with the rosemary and any oil left in the foil package. Add the remaining oil, the feta and lemon zest and gently toss to combine well. Season well with freshly ground black pepper and a little sea salt.

Divide the pastry in half. Roll out each half to 3 mm (⅛ inch) thick, then cut into eight 15 cm (6 inch) rounds. Put 2 tablespoons of filling on one half of each pastry round. Combine the egg yolk and milk in a small bowl and lightly brush the unfilled half of the pastry with the egg mix. Fold over the pastry to enclose the filling, pressing gently to seal well. Crimp the edge with your fingers, or gently press with the tines of a fork to seal. Repeat with the remaining filling and pastry rounds.

Place the pasties onto the tray, brush the tops with the remaining egg mixture, then put the tray in the refrigerator for at least 30 minutes. Remove and bake for 30 minutes, or until golden. Allow the pasties to cool a little before serving.

BAKED EGGPLANT (AUBERGINE) WITH TOMATO AND MOZZARELLA
SERVES 6

6 large slender eggplants (aubergines), cut in half lengthways, leaving stems attached
5 tablespoons olive oil
2 onions, finely chopped
2 garlic cloves, crushed
400 g (14 oz/1^2/$_3$ cups) tin chopped tomatoes
1 tablespoon tomato paste (concentrated purée)
3 tablespoons chopped flat-leaf (Italian) parsley
1 tablespoon chopped oregano
125 g (4 oz) grated mozzarella cheese

Preheat the oven to 180°C (350°F/Gas 4).

Score the eggplant flesh by cutting a criss-cross pattern with a sharp knife, being careful not to cut through the skin. Heat 2 tablespoons oil in a large frying pan, add 6 eggplant halves and cook for 2–3 minutes each side, or until the flesh is soft. Remove the eggplant, add 2 tablespoons oil to the pan and cook the remaining eggplant. Cool slightly and scoop out the flesh, leaving a 2 mm (1/8 inch) border. Finely chop the flesh, reserving the shells to create a bowl from the eggplant.

In the same pan, heat the remaining oil, add the onion and cook over medium heat for 5 minutes, or until softened. Add the garlic and cook for 30 seconds, then add the tomato, tomato paste, herbs and eggplant flesh. Cook over low heat, stirring occasionally, for 8–10 minutes, or until the sauce is thick and pulpy. Season well.

Arrange the eggplant shells in a lightly greased baking dish and spoon in the tomato filling. Sprinkle with mozzarella and bake for 5–10 minutes, or until the cheese has melted. Serve immediately.

LAMB STUFFED WITH OLIVES, FETA AND OREGANO
SERVES 4

85 g (3 oz/$^1/_2$ cup) kalamata olives, pitted
3 garlic cloves, crushed
100 ml (3$^1/_2$ fl oz) olive oil
800 g (1 lb 12 oz) lamb sirloin, trimmed (see Note)
90 g (3$^1/_4$ oz) feta cheese, crumbled
2 tablespoons oregano leaves, finely chopped
80 ml (2$^1/_2$ fl oz/$^1/_3$ cup) lemon juice
green salad, to serve

Combine the olives, garlic and 2 tablespoons of the olive oil in a food processor and process until smooth. Season to taste with freshly ground black pepper.

Prepare the sirloin by cutting horizontally most of the way through the piece, starting at one long side, taking care to not cut all the way through the end of the sirloin. Open out the lamb so you have a piece half as thick and twice as wide as you started with.

Spread the olive and garlic paste in a thin, even layer over the cut surface of the lamb, then crumble the feta over the top and scatter with the chopped oregano. Roll the lamb tightly, starting with one of the long cut edges, then tie the lamb with kitchen string to secure.

Put the lamb into a dish large enough to hold it lying flat. Drizzle with the lemon juice and remaining olive oil, turning to coat well. Cover and refrigerate for 3 hours.

Heat a chargrill plate or barbecue. Season the lamb to taste, then cook for 8–10 minutes over medium-high heat, turning once, or until it is cooked to your liking but still a little pink in the middle.

Transfer the lamb to a plate, cover loosely with foil and rest, covered, for 5 minutes. Cut off the string, then carve the lamb into 5 cm (2 inch) slices, on the diagonal. Serve immediately with a mixed green salad.

NOTE Use the thick end of the sirloin for this recipe.

Frisée, havarti and garlic crouton salad
SERVES 6–8

VINAIGRETTE
1 French shallot, finely chopped
1 tablespoon dijon mustard
60 ml (2 fl oz/¼ cup) sherry vinegar
170 ml (5½ fl oz/⅔ cup) extra virgin
  olive oil

SALAD
1 tablespoon olive oil
half a medium baguette, sliced
4 whole garlic cloves
1 baby frisée (curly endive), washed and
  dried
100 g (3½ oz/1 cup) walnut halves,
  toasted
90 g (3¼ oz/¾ cup) pitted green olives
200 g (7 oz) havarti cheese, cut into 1 cm
  (½ inch) pieces

For the vinaigrette, combine the shallot, mustard and vinegar in a bowl then slowly add the oil, whisking constantly until thickened. Set aside.

Heat the oil in a large frying pan, add the bread and garlic cloves and cook, tossing the pan often, over medium-high heat for 5–8 minutes, or until the croutons are crisp and golden. Remove from the heat and allow to cool. Break the croutons into small pieces. Discard the garlic.

Place the frisée, croutons, walnuts, olives, cheese and vinaigrette in a large bowl. Toss together well and serve.

Pesto bocconcini balls
SERVES 4–6

3 cups basil
40 g (1½ oz/¼ cup) pine nuts
35 g (1¼ oz/⅓ cup) grated parmesan
  cheese
2 garlic cloves
80 ml (2½ fl oz/⅓ cup) olive oil
300 g (10½ oz) bocconcini

Combine the basil, pine nuts, parmesan and garlic in a food processor and process until finely chopped. With the motor running, gradually add the olive oil and process until a paste is formed.

Transfer the pesto to a bowl and add the bocconcini. Mix gently to coat the bocconcini, cover with plastic wrap, then refrigerate for 2 hours before serving.

Frisée, havarti and garlic crouton salad

GOAT'S CHEESE GALETTE
SERVES 6

PASTRY
125 g (4½ oz/1 cup) plain (all-purpose) flour
60 ml (2 fl oz/¼ cup) olive oil

FILLING
1 tablespoon olive oil
2 onions, thinly sliced
1 teaspoon thyme leaves
125 g (4½ oz/½ cup) firm ricotta cheese (see Note)
100 g (3½ oz/¾ cup) crumbled soft goat's cheese
2 tablespoons pitted black olives
1 egg, lightly beaten
60 ml (2 fl oz/¼ cup) pouring (whipping) cream

For the pastry, sift the flour and a pinch of salt into a large bowl and make a well in the centre. Add the olive oil and mix with a flat-bladed knife until crumbly. Gradually add 3–4 tablespoons chilled water and mix until a coarse dough forms. Transfer to a lightly floured surface and form into a disc. Wrap in plastic wrap and refrigerate for 30 minutes.

For the filling, heat the olive oil in a frying pan, add the onion, cover and cook over low heat, stirring occasionally, for 30 minutes. Season to taste with sea salt and freshly ground black pepper, then stir in half the thyme. Cool slightly.

Preheat the oven to 180°C (350°F/Gas 4). Roll out the pastry on a sheet of baking paper to a 30 cm (12 inch) circle. Transfer the baking paper and pastry to a baking tray. Evenly spread the onion over the pastry, leaving a 3 cm (1¼ inch) border. Sprinkle the ricotta and the goat's cheese evenly over the onion. Place the olives over the cheeses, then sprinkle with the remaining thyme. Fold the pastry border in to the edge of the filling, gently pleating as you go.

Combine the egg and cream in a small bowl, then carefully pour over the filling. Bake in the lower half of the oven for 45 minutes, or until the pastry is golden. Serve warm or at room temperature.

NOTE Buy ricotta in bulk from a delicatessen or cheese shop; it has a better texture and fresher flavour than the bland, paste-like ricotta sold in tubs in supermarkets.

Baked prawns with feta
SERVES 4 AS A FIRST COURSE

300 g (10 oz) raw large prawns (shrimp)
2 tablespoons olive oil
2 small red onions, finely chopped
1 large garlic clove, crushed
350 g (12 oz) ripe tomatoes, diced
2 tablespoons lemon juice
2 tablespoons oregano, or 1 teaspoon dried
120 g (4¼ oz) feta cheese
extra virgin olive oil, for drizzling
chopped flat-leaf (Italian) parsley, to
  garnish

Preheat the oven to 180°F (350°F/Gas 4).

Peel the prawns, leaving the tails intact. Working from the head end, gently pull out and discard the intestinal tract (the dark vein) from each prawn tail.

Heat the olive oil in a saucepan over medium heat, add the onion and cook, stirring occasionally, for 3 minutes, or until softened. Add the garlic and cook for a few seconds, then add the tomato and cook for 10 minutes, or until the mixture is slightly reduced and thickened. Add the lemon juice and oregano, then season to taste.

Place the sauce into a shallow baking dish, about 15 cm (6 inches) square. Place the prawns on top. Crumble the feta over. Drizzle with oil and sprinkle with freshly ground black pepper.

Bake for 15 minutes, or until the prawns are just cooked. Serve immediately.

Cucumber, feta, mint and dill salad
Serves 4

120 g (4 oz) feta cheese
4 Lebanese (short) cucumbers
1 small red onion, thinly sliced
1 ½ tablespoons finely chopped dill
1 tablespoon dried mint
3 tablespoons olive oil
1 ½ tablespoons lemon juice

Crumble the feta into 1 cm (½ inch) pieces and place in a large bowl. Peel and seed the cucumbers and cut into 1 cm (½ inch) pieces, then add to the feta with the onion and dill, tossing to combine.

Grind the mint in a mortar using a pestle, or force through a sieve until powdered, then combine with the oil and juice in a bowl. Season to taste with salt and freshly ground black pepper then pour over the salad and toss well.

TOMATO AND BOCCONCINI SALAD
SERVES 4

3 large vine-ripened tomatoes
250 g (9 oz) bocconcini
12 basil leaves
60 ml (2 fl oz/¼ cup) extra virgin olive oil
1½ tablespoons balsamic vinegar, or to
  taste
sea salt

Slice each tomato into four 1 cm (½ inch) thick slices. Slice the bocconcini into twenty-four 1 cm (½ inch) thick slices. Arrange the tomato slices on a serving platter, alternating with 2 slices of the bocconcini and placing a basil leaf between the cheese and tomato slices. Drizzle the salad with olive oil and vinegar, season to taste with sea salt and freshly ground black pepper, then serve immediately.

MARINATED FETA
SERVES 6 AS PART OF AN ANTIPASTO PLATTER

350 g (11 oz) feta cheese
1 tablespoon dried oregano
1 tablespoon cracked black pepper
1 teaspoon coriander seeds
125 g (4 oz) sun-dried tomatoes in oil
4 small fresh red chillies
3–4 sprigs rosemary
olive oil

Dry the feta with paper towels, then cut into 2 cm (¾ inch) cubes. Place in a bowl and sprinkle with the oregano, pepper and coriander seeds.

Drain the sun-dried tomatoes over a bowl so you retain all of the oil. Arrange the feta, chillies, rosemary and sun-dried tomatoes in a sterilised 750 ml (24 fl oz/3 cup) wide-necked jar with a clip-top lid. Cover with the reserved sun-dried tomato oil — you should have about 60 ml (2 fl oz/¼ cup) — and top up with olive oil. Seal and refrigerate for 1 week before serving at room temperature.

Marinated feta will keep, refrigerated, for 1–2 months.

BAKED ONIONS STUFFED WITH GOAT'S CHEESE AND SUN-DRIED TOMATOES
SERVES 6

6 large white onions
60 ml (2 fl oz/¼ cup) extra virgin olive oil
1 garlic clove, crushed
80 g (2¾ oz/½ cup) sun-dried tomatoes, finely chopped
25 g (1 oz/⅓ cup) fresh white breadcrumbs
1 tablespoon chopped parsley
2 teaspoons chopped thyme
100 g (3½ oz) mild soft goat's cheese, or sheep's milk cheese, crumbled
80 g (2¾ oz/¾ cup) parmesan cheese, grated
1 egg
250 ml (9 fl oz/1 cup) vegetable or chicken stock
1 tablespoon butter

Preheat the oven to 180°C (350°F/ Gas 4).

Peel the onions, cut a slice off the top of each and reserve. Using a teaspoon, remove the centre of each onion, almost to the base, to create a cavity. Reserve the onion flesh for another use.

Cook the onion shells in a large saucepan of boiling water for 5 minutes, then drain and set aside.

Heat 2 tablespoons of oil in a small frying pan and cook the garlic for 1 minute, or until soft. Add the tomato, breadcrumbs and herbs and cook for 1 minute. Remove from the heat and add the goat's cheese and parmesan. Season to taste, add the egg and stir to combine well.

Divide the mixture among the onion cavities, pushing it in firmly. Place the onions in a large ovenproof ceramic dish. Pour the stock around the onions and drizzle with the remaining oil. Put the onion tops in place, cover the dish with foil and bake for 30 minutes, basting with the cooking juices from time to time. Remove the foil and cook for another 10 minutes.

Transfer the onions to a serving plate, then simmer the remaining stock over medium heat for 5–8 minutes, or until reduced by half and syrupy. Reduce the heat and whisk in the butter. The sauce should be smooth and glossy. Season to taste and spoon over the onions to serve.

VEAL INVOLTINI
SERVES 4

8 asparagus spears, trimmed
4 veal escalopes
4 thin slices mortadella (preferably with
    pistachio nuts)
4 thin slices bel paese cheese
plain (all-purpose) flour, seasoned with
    salt and pepper
3 tablespoons butter
1 tablespoon olive oil
3 tablespoons dry Marsala

Cook the asparagus in boiling salted water for 3 minutes, or until just tender. Drain well, reserving 60 ml (2 fl oz/¼ cup) of the cooking liquid.

Place each veal escalope between two sheets of plastic wrap and pound with a meat mallet to make four 12 x 18 cm (5 x 7 inch) rectangles. Season lightly with salt and freshly ground pepper. Trim both the mortadella and cheese slices to just a little smaller than the veal.

Cover each piece of veal with a slice of mortadella, then a slice of cheese. Place an asparagus spear in the centre, running across the width, with the tip protruding from the meat at one end. Place another asparagus spear alongside, but with its tip protruding from the other end. Roll each veal slice up tightly and tie each one twice with kitchen string to secure. Roll in the seasoned flour to coat, shaking off any excess.

Heat 2 tablespoons of the butter with the olive oil in a frying pan. Fry the rolls over low heat for about 10 minutes, turning frequently, until golden and tender. Transfer to a hot serving dish and keep warm.

Add the Marsala, the reserved asparagus liquid and the remaining butter to the pan and bring quickly to the boil. Simmer for 3–4 minutes, scraping up any pieces stuck to base of the pan, until the juices reduce and darken. Season to taste, then spoon the sauce over the veal rolls and serve immediately.

ZUCCHINI, THYME AND BOCCONCINI PIZZA
MAKES 2

PIZZA BASE

500 g (1 lb 2 oz/4 cups) plain
  (all-purpose) flour
7 g (1/8 oz/2 teaspoons) instant dried yeast
1 teaspoon sugar
1 tablespoon olive oil

8 zucchini (courgettes), cut into thin
  rounds
2 teaspoons grated lemon zest
15 g (1/2 oz/1/4 cup) finely chopped parsley
2 teaspoons thyme sprigs
4 garlic cloves, crushed
4 tablespoons olive oil
500 g (1 lb 2 oz) bocconcini cheese, finely
  diced
50 g (1 3/4 oz/1/2 cup) grated parmesan
  cheese
1 tablespoon extra virgin olive oil

To make the pizza base, combine the flour, yeast, sugar and 1 teaspoon salt in a large bowl, and make a well in the centre. Pour the oil and 310 ml (10 3/4 fl oz/1 1/4 cups) lukewarm water into the well and mix until the flour is incorporated and a soft dough forms. Turn onto a floured surface and knead for 10 minutes, or until the dough is smooth and elastic. Transfer to a greased bowl, turning to coat. Cover the bowl with plastic wrap and leave in a draught-free place for 40 minutes, or until doubled in size.

Preheat the oven to 220°C (425°F/Gas 7).

Gently knock the dough down and knead for 1 minute. Divide in half and roll each half out to a 5 mm (1/4 inch) thick round. Transfer the bases to two pizza trays.

Place the zucchini rounds, lemon zest, parsley, thyme, garlic and olive oil in a bowl and mix together.

Top each pizza base evenly with half the bocconcini and half the parmesan, then spoon on the zucchini mixture. Evenly distribute the remaining bocconcini and parmesan over the top, season well with salt and freshly ground black pepper, and drizzle with the extra virgin olive oil. Cook for 15–20 minutes, or until the base is crisp, and the topping is warmed through and golden.

SUPPLI
SERVES 6

1.5 litres (52 fl oz/6 cups) chicken stock
60 g (2¼ oz/¼ cup) butter
1 small onion, finely chopped
440 g (15½ oz/2 cups) arborio rice
75 g (2½ oz/¾ cup) grated parmesan
   cheese
sea salt
2 eggs, lightly beaten
9 basil leaves, torn in half
150 g (5½ oz) mozzarella cheese, cut into
   18 pieces about 1.5 cm (⅝ inch) square
150 g (5½ oz/1½ cups) dried
   breadcrumbs
vegetable oil, for deep-frying

Bring the stock to a simmer in a saucepan.

Melt the butter in a large saucepan, add the onion and cook over low heat, stirring often, for 3–4 minutes, or until softened but not browned. Add the rice and cook, stirring, for 1 minute, or until the rice is well coated and heated through. Add several ladlefuls of the hot stock to the rice, stirring continuously until the stock is absorbed, then add more stock and stir until absorbed. Continue adding stock and stirring until the rice mixture is creamy and the rice is *al dente*. Remove from the heat. Cool slightly, then add the parmesan and eggs. Season to taste with sea salt and freshly ground black pepper, then spread the mixture out onto a large baking tray to cool completely.

Divide the cooled rice into 18 even-sized portions. Take one portion and flatten it slightly across the palm of your hand, then place a piece of basil and a piece of mozzarella in the middle. Fold the rice around the cheese and basil to enclose then, using your hands, mould the rice into an egg-shaped croquette. Roll the croquette in breadcrumbs, pressing gently to coat, then place on a baking tray while you make the remaining croquettes.

Fill a deep-fryer or large saucepan one-third ful of oil and heat to 180°C (350°F), or until a cube of bread browns in 15 seconds when dropped into the oil. Deep-fry the suppli, in batches, for 4–5 minutes, or until golden, then drain on paper towel. Serve immediately.

GRATIN OF CREPES WITH PUMPKIN, GOAT'S CHEESE AND CRISP SAGE LEAVES
SERVE 6

CREPE BATTER
310 ml (10¾ fl oz/1¼ cups) milk
50 g (1¾ oz) butter, chopped, plus extra for pan-frying
155 g (5½ oz/1¼ cups) plain (all-purpose) flour
3 eggs

FILLING
400 g (14 oz) butternut pumpkin (squash), peeled
2 tablespoons olive oil
125 ml (4 fl oz/½ cup) vegetable oil
30 g (1 oz) sage leaves
250 g (9 oz/2 cups) crumbled soft goat's cheese

TOPPING
300 ml (10½ fl oz) pouring (whipping) cream
150 g (5½ oz) fontina cheese, grated

To make the crepe batter, gently heat the milk and butter in a small saucepan until the butter has melted; do not allow the milk to boil. Combine the flour and a pinch of salt in a large bowl and make a well in the centre. Add the eggs to the well and slowly whisk in the warm milk mixture. Whisk until completely smooth — the mixture will be quite thin. Cover and stand for 10–15 minutes.

Heat a nonstick frying pan over medium heat, then melt a little butter over the base. Add 60 ml (2 fl oz/¼ cup) of the crepe batter to the pan and swirl to cover the base. Cook for 30 seconds, or until the crepe is set and bubbles start to appear on the surface. Using a spatula, carefully turn the crepe over and cook for another 30 seconds. Transfer to a plate then repeat with the remaining batter until you have 12 perfect crepes. (There is enough batter to allow for a few practice crepes as well).

Heat a chargrill pan over medium heat. Cut the pumpkin into 24 slices, each about 1 cm (½ inch) thick. Put the pumpkin in a large bowl with the olive oil and ground pepper and toss to coat. Chargrill the pumpkin in batches for about 2 minutes, or until cooked through, turning once. Cool.

Heat the vegetable oil in a small frying pan until the oil creates a haze; take care that it does not burn. Add the sage leaves to the pan, in batches, cook until crisp, then remove and drain on paper towels.

Preheat the grill (broiler) to high. Place 2 pumpkin slices, some goat's cheese and a few sage leaves over one quarter of each crepe, saving some sage leaves to garnish. Fold each crepe in half, then in half again, to form neat triangles. Divide the crepes among four ovenproof gratin dishes or shallow pasta dishes.

To make the topping, heat the cream in a small saucepan, then stir in the grated cheese. Pour the mixture evenly over the crepes. Sit the dishes on a large baking tray, put the tray under the grill and cook for 3–5 minutes, or until the cheese is bubbling and hot. Scatter with the reserved crisp sage leaves and serve.

SMALL HAVARTI AND HONEY-CURED HAM SANDWICHES
MAKES 24

1 loaf of unsliced bread, sliced lengthways into 6 slices
125 g (4½ oz/½ cup) wholegrain mustard
100 g (3½ oz) shaved honey-cured ham
100 g (3½ oz) thinly sliced havarti cheese
55 g (2 oz/⅓ cup) mustard fruits, finely chopped (available from delicatessens)
40 g (1½ oz) butter
2 tablespoons olive oil

Preheat the oven to 120°C (235°F/Gas ½).

Brush each slice of bread with 1 tablespoon of mustard. Divide the ham and cheese among three of the bread slices, laying them on evenly, then sprinkle over the mustard fruits. Top each bread slice with one of the remaining slices of bread then cut eight rounds from each, using a 5 cm (2 inch) biscuit (cookie) cutter.

Melt half the butter and oil in a nonstick frying pan then, when the butter begins to foam, cook half the sandwiches until crisp and golden and the cheese is just starting to melt. Keep warm on a tray in a low oven while you cook the remaining sandwiches. Serve warm.

BROAD BEANS AND PASTA WITH SMOKED MOZZARELLA
SERVES 4

200 g (7 oz) broad (fava) beans
80 ml (2½ fl oz/⅓ cup) olive oil
2 garlic cloves, crushed
1 red capsicum (pepper), trimmed, seeded and finely shredded
a pinch of cayenne pepper
300 g (10½ oz) dried ditali, pennette or maccheroncini pasta
200 g (7 oz) smoked mozzarella cheese, cut into 1 cm (½ inch) pieces
3 tablespoons grated parmesan or grana padano cheese
sea salt

Cook the broad beans in a large saucepan of boiling, salted water for 2 minutes, or until tender. Drain well, reserving the cooking water. Transfer the beans to a bowl, cool and then peel.

Heat the oil in a large frying pan, add the garlic and capsicum and cook over low heat, stirring, for 3–4 minutes, or until the capsicum has softened. Stir in the cayenne pepper.

Meanwhile, bring the reserved water back to the boil, add the pasta and cook until al dente. Drain well, then add to the mixture in the pan along with the broad beans, mozzarella and parmesan. Toss to combine well, then gently heat until the mozzarella just begins to melt. Season to taste with sea salt and freshly ground black pepper, then serve immediately.

Small havarti and honey-cured ham sandwiches

semi-hard and hard cheeses

These two cheese families are large ones; their good keeping properties make them favourites of mass-producers and consumers alike. These are also versatile cheeses, with most being equally suited to the cheese board as they are to being cooked. There are dozens, if not hundreds of such cheeses, including cheshire, cantal, caerphilly, asiago, derby, leyden and, of course, cheddar.

There are three general categories of semi-hard cheeses — cheddar types (discussed below), washed-curd types, and the stretched-curd variety.

For washed-curd cheeses, the whey is mixed with water once the curd has formed and been cut. The hot water interrupts acid production, and when the cheeses are shaped the curds form easily into a dense, smooth-textured mass. Such cheeses, which include edam and gouda, have a soft, smooth texture and generally a mild flavour.

Stretched-curd, or *pasta filata*, cheeses are made similarly to washed-curd cheeses, but with hotter water; this makes the curds stretchy and pliable. They are then kneaded, cut into small pieces and pulled into long 'ropes', which expels moisture. The cheeses are aged for varying periods, their flavour changing depending upon the length of maturation.

Most semi-hard cheeses are pressed to some degree during manufacture to remove moisture; those that are aged lose even more moisture, becoming more pungent and firm as they do so. Semi-hard cheeses are excellent for the cook as they generally do not become greasy or rubbery when heated.

The hard cheese family represents cheeses that have the richest, most concentrated flavours of all cheeses, due to their age and low moisture content. These are the longest-keeping and largest of all the cheeses; size has much to do with their ability to slowly mature and to last a long time. Originally they were made in alpine regions of Europe as a way to use and preserve the excess milk from summer, but now they are made year-round. Curds for these cheeses are cooked and then cut, to get rid of as much moisture as possible; cooking sweetens the lactose as well, another characteristic of these cheeses. The curds are pressed to expel more moisture, and as a result, flavours are further concentrated. Grana padano and parmigiano reggiano are examples of hard cheese with a granular texture, while emmental is an example of a 'smooth' hard cheese (the curds are not cooked for as long, nor cut into such small pieces as for granular hard cheeses; the resulting cheese has a more elastic, buttery paste).

CHEDDAR

When we think of cheddar, we mostly think of the mass-produced, all-purpose stuff stocked by supermarkets. It may surprise some to know that, according to the Cheddar Presidia, recently formed under the auspices of the Slow Food Movement, there exist are only three cheeses that by rights can call themselves cheddar. The criteria, according to the Presidia, are that cheddar be made in Somerset, England, according to traditional practices — including the use of raw milk and animal, not vegetable, rennet — and be wrapped in cloth (the cloth binding helps the cheese to breathe). Unlike that of many other cheeses, the name 'cheddar' has never been legally protected, which is why much cheese is sold under that name.

Cheddar has been in production for a long time; records from 1170 show the English royal household purchasing quantities of the cheese, so it was clearly a well-established foodstuff by then. Cheddar has historically been held in the highest regard in England, and was not a daily staple for most people unless they were involved in the making of it. That began to change in the nineteenth century, when one Joseph Harding developed modern cheese-making processes whereby the curds could more easily be drained and the cheese mass-produced. Harding and his son Henry are held to be responsible for helping create the worldwide popularity of cheddar, teaching the Scottish, North Americans and Australians how to make it.

The vast majority of cheddar made today is a fairly bland, ultra-smooth-textured industrial product; it is worth seeking out genuine cheddar to experience the difference. Authentic cheddar has a complex set of flavours (including nutty, fruity, slightly sweet and tangy) and a hard, slightly crumbly paste. The texture is due to a special part of the processing called 'cheddaring'. After the milk has had rennet added to it, it is allowed to coagulate into a large mass. This is then cut into slabs, which are stacked on top of each other and allowed to drain. They are then restacked and allowed to drain again and this process is repeated until the curds are of a desired, uniform consistency, after which they are ground into very small pieces, salted and packed into cloth-lined moulds. The cheeses are then pressed for at least 12 hours. This process results in the subtly layered and crumbly texture that is such a characteristic of good cheddar.

Cheddar is aged in a cool environment for about 12 months, or longer for some very mature (or 'vintage') cheddars. Traditionally manufactured cheddars are hand turned during maturation so that they ripen evenly. A whole cheddar weighs up to 27 kg (60 lb); the large size helps ensure that the right balance of moisture loss and flavour and texture development can take place over the long maturing period. Colby cheese is similar to cheddar, although far milder in flavour and with a slightly different manufacturing process. Instead of the curds being cheddared, they are stirred in large vats of warmed whey before draining. It is thought that this process is an ancient one that has found favour with mass-producers as it is easy to mechanise. Club cheddar is fully mature cheddar that is ground and blended with other cheeses and, sometimes, other ingredients (such as herbs or dried tomatoes). This process changes the texture of the cheddar, allowing moisture to escape.

Cheddar has many uses in the kitchen — in gratins, sauces and savoury baked goods such as scones, pies and tarts. In parts of England it is even baked into apple pie. Use a good, mature cheddar for cooking, but save a 'true' English cheddar (such as Keen's, Montgomery's or Reade's) for the cheese board. Serve it with plum or quince paste, apples, dates, pears, grapes, dried muscatels or figs. The classic British ploughman's lunch sees cheddar partnered by chutney, pickled onions, crusty bread and ale or cider.

GRUYÈRE AND EMMENTAL

Although gruyère is most associated with Switzerland and takes its name from a village in the canton of Fribourg, the French also make gruyère-style cheeses, such as comte and beaufort. Graviera, a cheese popular in Greece, is also similar to gruyère in that it is firm-textured and has a sweet and nutty flavour. Gruyère is a hard, unpasteurised cow's milk cheese most closely associated with the classic Swiss dish fondue. The milk for making it must be processed 18 hours after it is collected; after curdling, the curds are cut into very small 'grains,' which are heated and then moulded. After the whey has drained out of the moulds, the cheeses are pressed and turned for 20 for 20 hours; after this they are soaked in a salty brine, drained again and matured in coolrooms. The flavour of gruyère changes greatly as the cheese matures. After five months, it is classified as 'mild' and has a creamy, nutty flavour. After eight to ten months, it is called 'semi-salted' and has a more assertive flavour. Finally, at twelve months, it is called 'reserve' and has a firm texture and complex, earthy flavours. The cheese develops a thin, brownish crust and internally is peppered with small holes. Gruyère is one of the best cheeses to use in cooking as it melts beautifully and its well-rounded flavour doesn't overwhelm other ingredients. It is used widely in quiches and pies, soups (such as French onion soup), toasted sandwiches (croque monsieur, for example), savoury breads and scones. It also makes a suburb table cheese.

Emmental cheese is the quintessential Swiss cheese for many people, with its shiny, pale yellow interior pierced with perfectly round, cherry-sized holes. The flavour of emmental is strong, fruity and a little nutty. The name emmental has never been legally protected, so cheeses by that name are made in various countries, from France to Germany and even Finland. True emmental, however, comes from the Emme Valley (*tal* means valley in German) in the Swiss canton of Bern, where it has been made since about the thirteenth century. A semi-hard cheese made from raw cow's milk (although many industrial versions use pasteurised milk), it is one of the largest cheeses; wheels can be up to 80 kg (175 lb), requiring 1000 litres (265 gallons) of milk.

Like gruyère, the cheese is curdled and cooked in copper vats then, after moulding, bathed in saline brine for 2 days to encourage a crust to form. The cheese then spends 6-8 weeks in a warm fermenting chamber during which time the bacteria used in the starter culture release carbon dioxide, forming the holes in the cheese. Acidity, temperature and time all affect the size of the holes, which can be as big as golf balls. The cheese is usually aged in cool cellars for eight to ten months (or up to 18 months to give it a fuller flavour). Emmental is used in fondues and is also used in New York's famous reuben sandwich. It is a perfect cooking cheese, but is just as delicious served with dark breads (such as rye or pumpernickel), cured meat, pickles and sauerkraut. It is also luscious with fruits and baguette or crackers. The very popular Norwegian cheese Jarlsberg is somewhat similar to Emmental and can be used in its place, especially in cooked dishes. As a table cheese, though, a good, aged Emmental is arguably superior.

FONTINA

Although many cheeses are marketed under the name, true fontina comes only from the Val d'Aosta region in Italy's far north, where a local breed of cow grazes on alpine pasture. The resulting milk is used, unskimmed and unpasteurised, within two hours of milking to make fontina. Historical references to fontina date back some 700 years, but it is believed to have been made even before then and has been hailed as one of Italy's (if not the world's) greatest cheeses. A semi-hard cheese, fontina is elegant in taste, with buttery, nutty, well-rounded flavours and a smooth, somewhat elastic, rich, straw-yellow paste with a few small holes throughout. It is aged in special caves carved from rock for about seven months, during which time it is salted, to inhibit the growth of mould, and brushed with liquid on alternate days, to keep the crust moist. Fontina is made in large wheels (up to about 20 kg/44 lb) and it has a thin, slightly concave crust that is a light red-brown in colour. Fontina originating from the Val d'Aosta has a green stamp on the rind that features the Alps. A distinguishing feature is fontina's ability to melt into a smooth, unctuous mass over low temperatures. For this reason the dish with which it is most closely associated is fonduta, a rich Italian version of fondue that includes butter, eggs and sometimes truffles. Fontina pairs well with cured meats such as salami, dense, flavoursome breads and fresh fruits.

GOUDA AND EDAM
Gouda is probably the most famous of the Dutch cheeses; in that country it accounts for about 60 per cent of all cheese produced. The name (actually pronounced *khow*-dah) comes from the town, just south of The Hague, where for centuries it has been sold at a large weekly market. The town's historic weighing house, dating from 1668, is still in use today and the weekly cheese weighing (in summer only) attracts large numbers of tourists. Gouda is a cow's milk cheese with a firm texture and a creamy taste that is unchanged since it was first made. To make gouda, cow's milk is treated with cultures then heated and the curds pressed into circular moulds. The young, shaped cheeses are soaked in brine to give a hard exterior and the characteristic slightly salty flavour. The cheeses are then coated with wax to prevent them from drying out as they age — aging times range from a few weeks to more than a year. Young gouda (coated in yellow wax) is smooth, buttery and mild in flavour, while aged examples (coated in black wax) are very firm and more piquant. These cheeses weigh 4½–11 kg (10–25 lb). 'Baby' gouda, coated in red wax, weighs about 500 g (1 lb). Gouda is also sold smoked and some varieties are flavoured with cumin seeds or herbs. Younger gouda can be used in cooking, and melts well. It is popular on the breakfast table in Europe, and makes a good addition to the cheese board, partnered with pickles or fruits, crackers or bread. Mature gouda, with its deep, complex flavours, is best savoured as a table cheese. Serve it with beer or red wine, fruits, cured meats or pickles and some good, crusty bread.

Edam is similiar to gouda; it is made from cow's milk, shaped in round moulds and sealed in a thin wax coating. Like gouda, edam is popular worldwide and today is made in many different countries, although it originates in the town of Edam in northern Holland. Edam is claimed to be the earliest exported cheese. Because it kept so well, the Dutch took it on their voyages of exploration, and later to their colonies, and thus its popularity spread early (between the fourteenth and eighteenth centuries). A major difference between the two is that edam is made using skim milk — farmhouse examples of edam have a fat content of 20–25 per cent, although factory-made edam is more likely to be about 40 per cent fat. Young edam has a very mild, slightly salty flavour; when aged, it is complex, piquant and prized by cheese lovers. Edam is made in rounds of 2.25–2.7 kg (5–6 lb), and in a 'baby' version of about 1 kg (2 lb 4 oz). It is usually sold at about 6 weeks of age, coated in a thin layer of red wax and tasting mild and faintly sweet. Edam that is 17 weeks or older has a black wax coating. Some edam is aged for nearly a year, when its texture is pleasantly hard and dry and its flavour a little salty. Edam can be used in cooking, in sandwiches or on a cheese board. Serve older edam with apples, pears and grapes and good ham or other cured meats. Younger edam pairs well with milder fruits, such as melon or cherries.

PROVOLONE

Provolone belongs equally in the 'semi-soft' family of cheeses as in the 'semi-hard'. There are two distinctly different styles of provolone made. One is called *provolone dolce* (or 'sweet' provolone). This cheese is semi-soft and mild, with a smooth, velvety texture and a pale yellow colour. The other type, *provolone piccante*, is aged for at least six to twelve months (sometimes as long as two years) and is semi-hard in texture and sharp tasting. As the cheese ages, the colour darkens. Both types are made from cow's milk and both are *pasta filata* cheeses. Another such cheese is mozzarella; to make *pasta filata* ('spun paste' in Italian) cheeses, the curds are kneaded while still hot; it is this step that gives these cheeses their characteristic elasticity. Apart from the aging time, the other key difference between the two types of provolone is the rennet used — *dolce* is made using calf's rennet while *piccante* uses goat or lambs' rennet. These different rennets affect flavour.

Provolone very closely resembles a southern Italian cheese called caciocavello, except provolone is made in the country's north. Confusingly, it used to be made in the south, in Campania. In the mid-nineteenth century, southern Italian cheese makers made their way north to an area around Cremona where, compared to what they had been used to near Naples, the pasture was lush and abundant. They continued making the same types of

cheese they always had (namely, provolone), which were quite different to
the parmesan native to that area of Italy; *provola* in the Neapolitan dialect
means 'globe-shaped'. Provolone is made using whole cow's or buffalo's milk
and comes in a variety of shapes and sizes. After the curds are kneaded and
worked, they are hand shaped into cheeses and aged. This shaping must be
done carefully so as to eliminate any pockets of air or liquid, and is a highly
skilled task. The cheeses are then immersed in a saline solution, wrapped and
tied at intervals with the strong twine that so characterises their appearance.
The cheese is then hung to age, during which time it acquires a thin, smooth
crust that is sometimes coated in wax for protection once aging is complete.
Provolone is sometimes smoked. It is a very versatile cheese, with many uses
in the kitchen. Either type can be served on a cheese board, and they are also
excellent cooked in such dishes as pasta, pizzas, salads, meatballs, risotto,
pies, polenta dishes and involtini (stuffed rolls of chicken, fish or meat).

RACLETTE

The word 'raclette' refers both to a specific cheese and a meal of the same
name, which is extremely popular winter fare in Switzerland and parts of
Germany and France. The cheese is an uncooked, cow's milk, semi-hard
cheese that originated in the hilly Valais region of Switzerland, although
it is also made in the Savoie and Franche-Comte regions of France (other
countries now make versions of raclette too). The meal called raclette
consists of the melted cheese served alongside boiled waxy potatoes and
pickles (cornichons and cocktail onions) and a sprinkling of paprika and
ground black pepper. Sometimes cured meats and sliced vegetables, such as
mushrooms and capsicum, are also served alongside, and the Swiss wine of
choice to wash it all down is a white from the Valais region called Fendant.
It is thought this meal originated in medieval times, when cow herders,
eating their simple meal of bread, cheese and wine in the chilly outdoors,
sat around a fire for warmth. Legend has it that someone once got too close
to the fire and in the course of doing so, melted the top layer of a piece of
cheese. Not wishing to waste the cheese, he scraped it onto a piece of bread,
ate it and pronounced it delicious — and the rest is history.

The word raclette comes from *racler*, which is the French verb meaning
'to scrape', and this is how cooked raclette is served. It is melted, in layers,
off a large piece of the cheese, and as the surface melts, it is scraped onto
a dining plate and the cheese returned to near the heat (traditionally an
open fire) for the surface to melt again. These days, electrical appliances
made especially for melting raclette are more commonly used than is a fire,
although purists insist that the only way to melt the cheese is over embers,

outdoors. Raclette cheese has a yellowish, hole-free paste, a light brown rind and a mild, nutty flavour reminiscent of gruyere; melting intensifies both the flavour and aroma. It is made in large (6 kg/13lb) rounds; sometimes smoked versions are made. Melted raclette has a particularly smooth, lush texture, which is a large part of the cheese's appeal when served this way. It is a wonderful, all-purpose cooking cheese and is also lovely served as a light meal with salads and bread, or fruits and bread in summer.

PARMESAN

'There was a huge amount of ground parmesan cheese and people were standing on it and the only thing they were doing was making macaroni and ravioli,' wrote Boccaccio in his *Decameron* in 1350. Today's producers of this most prized of all Italian cheeses claim it is made in exactly the same way today as it was in Bocaccio's day, yielding the same sweet, nutty aromas, rich, complex flavours and crumbly, granular texture that the famed Italian writer and his contemporaries would have enjoyed.

Parmesan, or at least a cheese very similar to it, existed at the beginning of the Christian era and it is believed to have been first produced in the Enza Valley, between the provinces of Parma and Reggio Emilia, in modern-day Emilia-Romagna in the north of Italy. It is still made in this area today and use of the name *parmigiano reggiano* is strictly protected. Genuine parmesan must come from a designated area around Reggiano, Modena and Parma, as must the grass and hay the cows eat in order to produce milk to make it. Parmesan is a hard, fine-grained cheese made from raw cow's milk, with small crystals distributed randomly through its paste, a result of the long ripening period (about two years) required to make authentic parmesan. Parmesan is a *grana* type of cheese; *grana*, the Italian word for 'grain', hints at its almost crunchy eating qualities and ability to grate exceptionally well.

'Parmesan' today is a generic term for a large variety of parmesan-type cheeses made globally, although in Emilia-Romagna, there are only about 600 artisanal producers who make the real thing. Mass-produced parmesan bears little resemblance to the prized *parmigiano reggiano*, with its delicate, lactic aromas and unmistakable sweet, mellow and nutty flavours. It is easy to identify 'real' parmesan, as each large wheel is marked on its thick, gold-yellow crust with a distinctive pattern of deep brown dots that spells 'parmigiano reggiano'. True parmesan is also marked by the formation of fine crystals in the paste, which look like very small white dots.

A whole wheel of parmesan weighs about 38 kg (84 lb) and requires some 570 litres (150 gallons) of milk. To make parmesan, milk from the evening milking is skimmed. This milk is combined with unskimmed milk from the next (the morning) milking then poured into enormous copper cauldrons; most other cheeses are made using stainless steel. The milk is bought to a temperature of 32–33°C (90–91°F) and stirred as rennet is added; each cauldron holds about 1000 litres (265 gallons) of milk and produces 60–70 kg (132–154 lb) of curds. The curds are cut into very, very tiny pieces; machinery is sometimes used to do this, although some traditional producers still use a special hand-cutter for the job. The mixture is heated again then the curds are collected into large linen cloths, moulded and weighted to eliminate any remaining whey. The cheeses are turned in the moulds for a time to allow a crust to form and then are submerged in a saline solution where they are left and turned regularly, for some 20 days. After this the cheeses are stored on wooden boards in special large, airy rooms where they are left to age for up to two years; there are some parmesans, called *stravecchio*, that are aged for more than three years.

After the first eight months, the cheeses will be inspected by being tapped with a hammer-like device. The sound it makes should be uniform; an uneven tone signifies holes or fissures in the interior and the cheese will be given an inferior grading and not sold as *reggiano*. During this time the cheeses are all turned regularly; more frequently to begin with, and in the latter stages every 10–12 days. It is over this length of time that the cheese acquires its unique, nutty, sweet flavours and hard, crystalline texture. The size of the cheese protects it from drying out during this long period and the long curing time also explains the expense of the cheese, as it is costly to hold it for this amount of time.

Culinarily, parmesan has many uses. It is an essential ingredient in many Italian dishes: pastas (both in fillings for stuffed pastas and sprinkled over pasta about to be served); egg dishes, such as frittata and omelettes; stuffings for vegetables and meats; pizzas, pies and foccaccia; and shaved over *carpaccio* (paper-thin slices of raw beef fillet) and salads. The true test of a fine parmesan is to eat it crumbled into pieces, with such simple accompaniments as aged balsamic vinegar, cubes of good mortadella, slices of prosciutto, olives or fruits, such as grapes or pears. Inferior parmesan will be uninteresting eaten thus; the complexity of true *reggiano*, on the other hand, will shine. Parmesan has a fat content of 25–28 per cent and is low in moisture. Once a piece is cut from the whole round, parmesan will keep, refrigerated, for up to a month; wrap it in waxed paper and then foil, replacing the paper after every time you use the cheese. Grated parmesan can be frozen, if necessary, although its flavour will be a little diminished. Use any frozen grated parmesan stirred into soups, pastas or risottos.

PECORINO

Pecorino is not just one cheese; it is a fascinating family of similar, regional cheeses from Italy. There are about a dozen distinct types in Italy, with *pecorino Toscano*, *pecorino Sardo*, *pecorino Romano* and *pecorino Siciliana* being the best known. Made using sheep's milk, varying in flavour from mild to sharp and salty, and spanning semi-hard to hard textures (depending upon age), pecorino is a most appealing and versatile cheese. Aged versions bear some similarity to parmesan, but pecorino is more earthy in character.

Pecorino cheeses have very long histories; *pecorino romano* originated in Roman times. To make it, sheep's rennet is used as the coagulant for the sheep's milk and once the cheeses are formed, they are rubbed all over with salt, pierced all over so the salt can penetrate the interior, then left to age for about eight months. *Pecorino romano* is grated over vegetable dishes and pasta, and its strong flavour makes it a natural accompaniment for assertive, slightly bitter salad vegetables, such as radishes, rocket and watercress.

The Tuscan version of pecorino tends, even in its most mature form, to be milder than those from other regions of Italy. The Tuscans have been making such cheese for at least 2000 years — Pliny the Elder makes reference to the *pecorino Toscano* trade in his book *Natural History*, which dates from the first century AD. When young, this pecorino is semi-hard and white in colour and is appreciated as a table cheese. When well aged, it takes on a yellow colour, a hard texture and sharper flavour and is mainly used grated over hearty soups and pasta dishes; sometimes the Tuscans eat it with bread, drizzled with honey. This pecorino, as are those from other regions, is often aged using various media that affect flavour; for example, it can be wrapped in sack cloth and buried for 100 days (the resulting cheese is called *pecorino di fossa* and it has a very special flavour), wrapped in chestnut leaves and aged (*pecorino castagno*) or rubbed all over with the skins and seeds left over when grapes are pressed for wine (*pecorino vinacce*).

The Sicilians farm more sheep than any region in Italy except Sardinia, and pecorino is a cheese staple over the whole island. Their cheese lays claim to a mention in Homer's *Odyssey*, which dates it to as far back as 800 BC. Sicilians often flavour their pecorino with black pepper or dried chilli flakes for extra pungency. They call young, unsalted pecorino *tuma*, and use this soft, rather delicate cheese as a table cheese. The next stage of maturing involves salting and aging the cheese for four months, after which time it is still somewhat soft and mild; a second salting and further aging produce a hard cheese for grating and use in cooked dishes. Sardinian pecorino (*pecorino sardo*) can be milder than its ancient romano counterpart when young but stronger and spicier when older. It uses sheep's milk but calves' rennet and is dry salted during aging. After just two months' aging it becomes somewhat spicy flavoured and is an excellent grating cheese for use in pasta and vegetable dishes. Young Sardinian pecorino is sweet and delicate and perfect as a table cheese, in vegetable dishes and in salads.

GRANA PADANO

Another *grana* type of Italian hard cheese (*grana* means 'grain' and refers to its texture), grana padano is extremely similar to parmesan and is often confused with it. There are, however, some key differences. The main one is the feed; the cows that produce grana eat silage and other fodder such as alfalfa and maize, as well as the hay and grass that go into the milk for parmesan. Its texture is indeed grainier than parmesan, with small, crumbly fissures radiating through the paste. In colour it tends to be a little paler than parmesan. A hard, cooked-curd cheese, grana padano is made from semi-skimmed pasteurised milk and is matured slowly in large wheels; it is at its prime after 15–18 months of aging, compared to the 20–24 months required for parmesan. These days grana padano is perceived as being more of an industrialised product than parmesan. It is still, however, a sensational and highly versatile cheese, and one with a venerable history. It was purportedly first made by Cistercian monks around the twelfth century as a way of preserving excess milk, and by 1477 was supposedly one of Italy's most famous cheeses. It keeps well (up to two years with proper storage) and quickly became a cherished food of both aristocrats and peasants alike. Today it is made by many producers in Italy's north, notably in the regions of Lombardy, Emilia-Romagna, Trentino, Piedmont and Veneto — the word *padano* refers to the river Po, which runs though many of these regions. With its sweet–savoury, buttery flavours and appealingly crumbly texture, grana padano is a very useful cheese, both as a table cheese with simple accompaniments (pears or apples, cured meats such as salami, olives and preserved vegetables) and grated into cooked dishes containing rice, polenta, pasta or vegetables. As with parmesan or any other hard cheese, do not buy pre-grated grana padano, as its flavour and aromas are quickly lost. It keeps for a long time, so buy a large chunk and grate off only as much as you need.

Many countries make versions of grana padano, but there are ways to ensure you buy the genuine Italian cheese. Firstly, look at the rind. Not only should it be smooth, thick and slightly oily, but if it has a four-leaf clover symbol etched into it, this means it is D.O.P. certified and is the 'real thing'. Good grana padano will be a pale straw colour in the centre, intensifying near the rind. For best keeping results, wrap your grana in a slightly moistened cloth and refrigerate it at about 4.5°C (40°F).

FENNEL FRITTERS
SERVES 4

1 kg (2 lb 4 oz) fennel bulbs, tough outer
  leaves removed
30 g (1 oz/^{1}/$_3$ cup) grated pecorino cheese
80 g (2^3/$_4$ oz/1 cup) fresh breadcrumbs
60 g (2 oz/^{1}/$_2$ cup) plain (all-purpose) flour
3 eggs, lightly beaten
olive oil, for shallow-frying
lemon wedges, to serve

Trim the base of the fennel and remove any small stalks. Slice the fennel lengthways into 5 mm (^{1}/$_4$ inch) widths then cook in boiling salted water for 3 minutes, or until tender. Drain and pat dry. Leave to cool.

Combine the cheese and breadcrumbs in a bowl and season with salt and freshly ground pepper.

Dust the fennel in the flour, shake off the excess then dip in the beaten egg, draining off excess. Coat the fennel in the crumb mixture then heat the oil in large heavy-based frying pan until the oil begins to sizzle. Add the fennel to the pan in batches, being careful not to overcrowd the pan, then cook for 2–3 minutes each side, or until golden brown and crisp. Drain on paper towels, season to taste and serve accompanied with the lemon wedges.

WARM CIDER AND RACLETTE DIP
MAKES 500 G (1 LB 2 OZ/2 CUPS); SERVES 6

40 g (1^1/$_2$ oz) unsalted butter
3 French shallots, finely chopped
125ml (4 fl oz/^{1}/$_2$ cup) sweet cider
300 g (10^1/$_2$ oz/ about 2^1/$_4$ cups) chopped
  raclette cheese
toasted sourdough raisin bread, to serve

Melt the butter in a saucepan, add the shallots and cook over low heat, stirring frequently, for 6–8 minutes, or until soft. Stir in the cider then heat gently until warm. Add the raclette and stir constantly until the cheese melts and the mixture is glossy and smooth; do not allow the mixture to boil. Transfer to a bowl then serve immediately, with toasted raisin bread passed separately.

Asparagus, pecorino and mint frittata
SERVES 4

6 eggs
120 g (4½ oz/1⅓ cups) grated pecorino
  cheese
1 large handful mint, finely shredded
200 g (7 oz) baby asparagus spears,
  trimmed
2 tablespoons extra virgin oil
sea salt

Put the eggs in a large bowl, whisk well, then stir in the cheese and mint and set aside. Preheat the oven grill (broiler) to medium-high.

Cut the asparagus diagonally into 5 cm (2 inch) pieces. Heat the oil in a 20 cm (8 inch) frying pan. Add the asparagus and cook for 4–5 minutes, stirring often, until tender and bright green. Season with sea salt and freshly ground black pepper to taste, then reduce the heat to low.

Pour the egg mixture over the asparagus and cook for 8–10 minutes, using a spatula to gently pull the sides of the frittata away from the sides of the pan and tipping the pan slightly so the egg runs underneath the cooked base.

When the mixture is nearly set but still slightly runny on top, place the pan under the grill (broiler) for 1–2 minutes, until the top is set and just browned. Serve warm or at room temperature, cut into wedges.

SWISS-STYLE CHICKEN
SERVES 4

4 x 200 g (7 oz) skinless, boneless chicken
   breasts
1 tablespoon extra virgin olive oil
40 g (1½ oz) butter
1 garlic clove, crushed
200 g (7 oz) button mushrooms, sliced
1 tablespoon chopped tarragon
125 ml (4 fl oz/½ cup) pouring
   (whipping) cream
1 tablespoon brandy
4 large slices gruyère or Swiss cheese

Heat the oven grill (broiler) to high. Place the chicken breasts between two sheets of plastic wrap and gently pound with a meat mallet until each fillet is 1 cm (½ inch) thick.

Place the chicken breasts on a lightly oiled grill tray and brush them with the oil. Grill for 6–8 minutes, turning once, or until cooked through. Transfer to a lightly oiled shallow ovenproof dish.

While the chicken is grilling, melt the butter in a frying pan. Add the garlic and mushrooms and cook over medium heat for 3 minutes, or until the mushrooms have softened. Add the tarragon, cream and brandy and stir over high heat for about 2 minutes, or until the sauce has reduced and thickened.

Spoon the hot sauce over the chicken breasts and top each with a slice of cheese. Put the dish under the hot grill and cook for about 4 minutes, or until the cheese has melted. Serve immediately.

Parmesan wafers
Makes 30

125 g (4½ oz/1¼ cups) grated parmesan cheese
1 tablespoon plain (all-purpose) flour
2 tablespoons thyme

Preheat the oven to 220°C (425°F/Gas 7). Line two baking trays with baking paper and, using a 7 cm (2¾ inch) biscuit (cookie) cutter as a guide, draw circles on the paper. Turn the paper upside-down on the trays.

Toss the cheese and flour together in a bowl, then sprinkle 2 teaspoons of the mixture over 3–4 circles on the paper, spreading the mixture to the edge of each round. Scatter a few thyme leaves over each round.

Bake in batches for about 3 minutes, or until the cheese is melting but not hard. Using a spatula, turn the rounds over and cook for a minute more, or until firm and light golden. Remove each round from the tray and drape over a rolling pin or bottle until cool. Repeat with the remaining mixture.

Serve the wafers at room temperature on the day they are made.

Cauliflower and smoked cheese rarebit
Serves 4

½ loaf ciabatta, cut into 8 thick slices
1 garlic clove
800 g (1 lb 12 oz) cauliflower, cut into
  small florets
120 g (4¼ oz/1 cup) grated gruyère or
  emmental cheese
120 g (4¼ oz/1 cup) grated smoked
  cheddar cheese
1 tablespoon dijon mustard
2 eggs, beaten
2 tablespoons beer
4 tablespoons pouring (whipping) cream

Heat the grill (broiler) and toast the ciabatta on both sides. Cut the garlic clove in half and rub the cut sides over one side of each slice of ciabatta.

Bring a saucepan of water to the boil and cook the cauliflower for about 5 minutes, or until tender when pierced with a knife. Drain well.

Combine the cheeses, mustard, egg, beer and cream in a bowl and whisk to combine well. Put the toast on a baking tray and divide the cauliflower among the toast. Spoon the cheese mixture over, taking care to coat the cauliflower well.

Place the toast under the grill and cook until the cheese mixture is golden brown and bubbling. Serve immediately.

TOASTED CHEESE, AÏOLI AND PASTRAMI SANDWICH
SERVES 4

1 loaf ciabatta or Turkish bread
2 garlic cloves, crushed
125 g (4½ oz/½ cup) whole-egg
    mayonnaise
8 pastrami slices
100 g (3½ oz) semi-dried (sun-blushed)
    tomatoes, chopped
2 tablespoons capers, rinsed, drained and
    chopped
6–8 slices cheddar cheese

Heat the grill (broiler). Cut the loaf in half horizontally and then into four even-sized pieces. Toast all the pieces under the grill.

To make the aïoli, stir the garlic into the mayonnaise and season well with salt and pepper.

Spread the aïoli over the cut sides of bread. Place a slice of ham on four of the pieces and then divide the semi-dried tomatoes and capers among them. Top with cheese to cover, then place on a baking tray.

Grill the sandwiches until the cheese melts and starts to bubble, then top with remaining toast pieces and press firmly.

Cut each sandwich in half diagonally and serve at once.

TWICE-BAKED CHEESE SOUFFLÉS
SERVES 4

250 ml (9 fl oz/1 cup) milk

3 whole black peppercorns

1 onion, cut in half and studded with
   2 cloves

1 bay leaf

60 g (2¼ oz/¼ cup) butter

60 g (2¼ oz/½ cup) self-raising flour

2 eggs, separated

125 g (4½ oz/I cup) grated gruyère
   cheese

250 ml (9 fl oz/1 cup) pouring (whipping)
   cream

50 g (1¾ oz/½ cup) finely grated
   parmesan cheese

Preheat the oven to 180°C (350°F/Gas 4). Lightly grease four 125 ml (4 fl oz/½ cup) ramekins.

Place the milk, peppercorns, onion and bay leaf in a saucepan and heat until nearly boiling. Remove from the heat and let the milk infuse for 10 minutes. Strain, discarding the solids.

Melt the butter in a saucepan, add the flour and cook over medium heat for 1 minute. Stirring constantly, add the milk a little at a time, returning the mixture to a simmer between additions and stirring well to prevent lumps from forming. Simmer, stirring, until the mixture boils and thickens.

Transfer the mixture to a bowl, add the egg yolks and gruyère cheese and stir to combine well.

Using electric beaters, whisk the egg whites until soft peaks form, then gently fold into the cheese sauce. Divide the mixture among the ramekins and place in a baking dish half-filled with hot water. Bake for 15 minutes. Remove from the baking dish, cool and refrigerate until needed.

Preheat the oven to 200°C (400°F/Gas 6), remove the soufflés from the ramekins and place onto ovenproof plates. Pour the cream over the top and sprinkle with parmesan. Bake for 20 minutes, or until puffed and golden. Serve immediately.

ALIGOT
SERVES 6–8

800 g (1 lb 12 oz) floury potatoes, cut into chunks
75 g (2¹/₂ oz) butter
2 garlic cloves, crushed
3 tablespoons milk
300 g (10¹/₂ oz/2¹/₂ cups) grated cantal cheese (or mild cheddar)

Cook the potatoes in boiling salted water for 20–30 minutes, or until tender.

Meanwhile, melt the butter in a small saucepan over low heat and add the garlic. Mash the potatoes and then pass them through a sieve to give a really smooth purée (don't use a food processor, or they will become gluey).

Return the potato purée to the saucepan over gentle heat and add the garlic butter and milk. Mix together well then beat in the cheese, handful by handful—once it has melted, the mixture will be stretchy. Season with salt and freshly ground black pepper before serving.

CHEESY BUBBLE AND SQUEAK CAKES WITH BACON
SERVES 4

4 large or 8 small roasting potatoes (about 1 kg/2 lb 4 oz)
2 tablespoons milk
2 tablespoons butter
480 g (1 lb/6¹/₂ cups) finely shredded savoy cabbage
120 g (4¹/₄ oz/1 cup) grated cheddar cheese
1 tablespoon oil
8 bacon slices

Cut the potatoes into pieces and cook them in boiling salted water for 15 minutes, or until tender. Drain well, return them to the pan with the milk and mash until they are smooth. Season to taste with salt and freshly ground black pepper.

Melt the butter in a nonstick frying pan, add the cabbage and cook, stirring, for about 5 minutes, or until soft. Add the cabbage and cheese to the potato mixture and form into cakes; make either small ones or larger ones, as preferred.

Heat the oil in the same frying pan over medium heat, add the bacon and cook for 5–6 minutes, turning once, or until crisp. Remove the bacon from the pan and keep warm. Add the potato cakes to the pan and fry them for 3–5 minutes on each side, depending on size, until they are well browned and slightly crisp. Shake the pan occasionally to move the cakes around so they don't stick. Serve with the bacon.

CUMIN AND GOUDA GOUGÈRES
MAKES 40

100 g (3½ oz) butter, chopped
140 g (5 oz) plain (all-purpose) flour
½ teaspoon cumin seeds, lightly crushed
3 eggs
150 g (5½ oz/1¼ cups) finely grated aged
 gouda cheese

Preheat the oven to 200°C (400°F/Gas 6). Line a baking tray with baking paper.

Heat 250 ml (9 fl oz/1 cup) water, the butter and ¼ teaspoon salt in a small saucepan over medium heat until the butter has melted and the mixture has just come to the boil. Add the flour and cumin and stir until the mixture comes away from the side of the saucepan.

Transfer the mixture to the bowl of an electric beater and allow to cool a little (alternatively, use a hand mixer or wooden spoon). Beating continuously, add the eggs one at a time, beating well after each addition, until the mixture is glossy and drops heavily from the beaters. You may not need all the egg, and take care not to add too much egg. Stir in the cheese.

Put teaspoonfuls of the mixture, about 4 cm (1½ inches) apart, on the baking tray. Bake for 20 minutes, then reduce the heat to 160°C (315°F/Gas 2–3) and cook for a further 20 minutes, or until the gougères are puffed, golden and dry. Turn off the oven, open the door slightly and leave the gougères to cool a little. Serve warm or at room temperature with drinks.

Shallot, bacon and cheddar muffins

Makes 6

60 ml (2 fl oz/1/4 cup) oil, plus
  2 teaspoons, extra
5 shallots, peeled
2 bacon slices, finely chopped
250 g (9 oz/2 cups) plain (all-purpose)
  flour
1 tablespoon baking powder
1 tablespoon caster (superfine) sugar
  1 teaspoon dry mustard
140 g (5 oz/1 1/4 cups) grated mature
  cheddar cheese
185 ml (6 fl oz/3/4 cup) milk
1 egg
sweet paprika, to serve

Preheat the oven to 200°C (400°F/Gas 6). Grease a 6-hole giant muffin tin. Slice 1 of the shallots into rings. Heat 2 teaspoons of oil in a frying pan over low heat, add the shallot and fry for 3 minutes or until softened, then remove and drain on paper towels. Set aside.

Finely chop the remaining 4 shallots. Increase the pan heat to medium, add the shallots and bacon to the pan, then cook for 5 minutes, or until the shallots are soft. Drain on paper towels.

Sift the flour, baking powder, sugar, mustard and 1/2 teaspoon salt into a bowl. Add 90 g (3 1/4 oz/3/4 cup) of the cheddar cheese and the bacon and shallots and stir to combine well. Combine the milk, egg and remaining oil in a jug. Pour into the flour mixture and stir quickly and gently to combine; the batter should be a little lumpy. Divide the batter among the muffin holes. Top the muffins with the fried shallot and remaining cheddar cheese. Bake for 20–25 minutes, or until the muffins are golden and a cake tester inserted into the centre of a muffin comes out clean. Cool in the tin for 5 minutes, then turn out onto a wire rack to cool completely. Sprinkle paprika on top to serve.

OPEN LASAGNE WITH ROCKET AND WALNUT PESTO
SERVES 4

PESTO
100 g (3½ oz/1 cup) walnuts
2 garlic cloves
2 large handfuls baby rocket (arugula)
1 large handful basil
1 large handful flat-leaf (Italian) parsley
100 ml (3½ fl oz) extra virgin olive oil
80 ml (2½ fl oz/⅓ cup) walnut oil
50 g (1¾ oz/½ cup) grated pecorino cheese
100 g (3½ oz/1 cup) grated parmesan cheese

375 g (13 oz) fresh lasagne sheets
1 tablespoon olive oil
100 g (3½ oz/2 cups) baby English spinach
1 garlic clove, sliced
2 tablespoons lemon juice
200 g (7 oz/1⅔ cups) crumbled marinated goat's feta cheese
2 tablespoons finely grated parmesan cheese

To make the pesto, preheat the oven to 180°C (350°F/Gas 4). Rinse the walnuts in cold water, shake dry, spread on a baking tray and bake for 5–8 minutes, or until light golden.

Transfer the walnuts to a food processor and add the garlic, rocket, basil and parsley. Using the pulse button, process the mixture just until it resembles coarse breadcrumbs. With the motor running, add the oils in a thin stream, then add the pecorino and parmesan and process for 40 seconds. Transfer the pesto to a bowl, cover with plastic wrap and set aside until needed.

Cut the lasagne sheets into sixteen 8 cm (3¼ inch) squares. Cook a few squares at a time in a large saucepan of boiling salted water for 4 minutes, or until *al dente*. Lay them on a clean tea towel (dish towel) and cover to keep warm while the remaining squares cook.

Heat the olive oil in a large frying pan over medium heat, add the spinach and garlic and sauté until just wilted. Stir in the lemon juice, cover and keep warm.

Spoon 1 tablespoon of the pesto onto four warmed plates and spread out with the back of the spoon to the size of one of the pasta squares. Cover with a pasta square, then divide one-third of the spinach evenly among the plates, on top of the pasta. Sprinkle one-third of the goat's feta evenly among the plates, cover with another pasta square and spread with the pesto. Repeat the layers, finishing with a layer of pesto. Sprinkle with the grated parmesan and serve immediately.

ARTICHOKE AND PROVOLONE QUICHES
MAKES 6

250 g (9 oz/2 cups) plain (all-purpose)
  flour
125 g (4$^1/_2$ oz/$^1/_2$ cup) butter, cut into
  cubes
1 egg yolk
about 3 tablespoons iced water

FILLING
6 eggs, lightly beaten
3 teaspoons wholegrain mustard
200 g (7 oz/1$^1/_2$ cups) grated provolone
  piccante cheese
300 g (10$^1/_2$ oz) marinated artichokes,
  sliced
125 g (4$^1/_2$ oz/$^3/_4$ cup) chopped semi-dried
  (sun-blushed) tomatoes

To make the pastry, combine the flour and butter in a bowl then, using your fingertips, rub in the butter until the mixture resembles breadcrumbs. Make a well in the centre, add the egg yolk and 3 tablespoons iced water to the well then stir with a flat-bladed knife until a coarse dough forms, adding a little more water if needed. Turn out onto a floured surface and gather into a ball. Wrap in plastic wrap and refrigerate for at least 30 minutes.

Preheat the oven to 190°C (375°F/Gas 5). Grease six 11 cm (4$^1/_2$ inch) fluted pie tins.

To make the filling, combine the eggs, mustard and cheese in a bowl.

Divide the pastry into six even-sized pieces. Roll each out to a circle about 3 mm ($^1/_8$ inch) thick and use it to line the prepared tins. Trim the edges of the pastry. Divide the artichokes and tomatoes among the pastry cases, pour the egg mixture over and bake for 25 minutes, or until golden.

GRILLED POLENTA WITH SHAVED FENNEL SALAD
SERVES 6

500 ml (17 fl oz/2 cups) milk
175 g (6 oz) polenta
35 g (1¼ oz/⅓ cup) grated grana padano
   cheese, plus 2 tablespoons extra, to
   serve
1 tablespoon butter
200 g (7 oz) fennel bulb, trimmed
60 g (2¼ oz/2 cups) watercress leaves
1 tablespoon lemon juice
2 tablespoons olive oil

In a heavy-based saucepan, bring the milk and 500 ml (17 fl oz/2 cups) of water to the boil. Add the polenta, whisk to combine well, then reduce the heat to very low and cook for 40 minutes, stirring often to prevent it from sticking. Remove from the heat, stir in the grated cheese and butter and season to taste with salt and freshly ground black pepper. Pour into a well-greased tray, form into a circle about 2 cm (¾ inch) thick and leave to cool.

Cut the polenta into six wedges. Heat a hot chargrill pan or a barbecue grill plate, brush with a little olive oil, then cook the polenta for about 3 minutes on each side or until crisp, golden and well-marked from the grill.

Slice the fennel as thinly as possible and chop the fronds. Toss in a bowl with the watercress, lemon juice, oil and half the shaved cheese. Season to taste.

Serve the chargrilled polenta with the fennel salad piled to one side, and the remaining shaved cheese on top.

PASTICCIO OF MACARONI AND RAGU
SERVES 6

FILLING
40 g (1½ oz) butter
1 tablespoon olive oil
1 onion, finely chopped
2 garlic cloves, crushed
500 g (1 lb 2 oz) minced (ground) beef
60 g (2¼ oz/⅔ cup) finely sliced button
  mushrooms
115 g (4 oz) chicken livers, trimmed and
  finely chopped
a pinch of nutmeg
1 tablespoon dry Marsala
3 tablespoons dry white wine
2 tablespoons tomato paste (concentrated
  purée)
250 ml (9 fl oz/1 cup) chicken stock
3 tablespoons grated parmesan cheese
1 egg

PASTRY
300 g (10½ oz) plain (all-purpose) flour
2 teaspoons caster (superfine) sugar
100 g (3½ oz) cold butter, cubed
1 egg
about 2 tablespoons chilled water

150 g (5½ oz) macaroni
100 g (3½ oz) ricotta cheese
50 g (1¾ oz/½ cup) grated Parmesan
  cheese
2 tablespoons milk
a pinch of cayenne pepper
1 egg, beaten
120 g (4¼ oz) grated mozzarella cheese

To make the filling, heat the butter and oil in a large frying pan and cook the onion and garlic over medium heat for 5–6 minutes, or until golden. Add the beef, increase the heat to high and cook until browned, breaking up any lumps with a wooden spoon.

Add the mushrooms, liver and nutmeg, season and cook until the liver changes colour. Add the Marsala and wine and cook until they evaporate. Stir in the tomato paste and stock, reduce the heat and simmer for 30 minutes. Remove from the heat and season. Beat the parmesan and egg together and quickly stir through the meat.

To make the pastry, sift the flour, sugar and 1 teaspoon salt into a bowl and rub in the butter until the mixture resembles coarse breadcrumbs. Add the egg and the cold water, a little at a time, stirring until the dough comes together. Transfer to a lightly floured board and knead lightly until smooth. Cover with plastic wrap and refrigerate until needed.

Cook the macaroni in a large saucepan of boiling salted water until *al dente*, then drain.

Meanwhile, combine the ricotta, parmesan, milk, cayenne and half the egg in a bowl and stir until smooth. Season well and stir in the macaroni.

Preheat the oven to 180°C (350°F/Gas 4). Grease a deep-sided pie dish.

Divide the pastry into two balls, one larger than the other. On a lightly floured work surface, roll out the larger ball to line the dish, allowing overlapping edges. Roll out the second ball large enough to cover the pie dish. Spread half the filling into the pastry-lined pie dish, then layer half the pasta over it. Top with half the mozzarella. Repeat the layers.

Brush the pie rim with beaten egg and cover with the pastry lid. Brush the top with egg and make three slits in the centre with a sharp knife. Bake for 50 minutes, or until golden. Leave to rest for 10 minutes before serving.

FILLED CHEESE BISCUITS

SERVES 4–6

BISCUIT PASTRY
125 g (4½ oz/1 cup) plain (all-purpose)
    flour
½ teaspoon baking powder
60 g (2½ oz/¼ cup) cold butter, chopped
1 egg, lightly beaten
60 g (2¼ oz) grated cheddar cheese
1 teaspoon finely snipped chives
1 teaspoon finely chopped marjoram
1 tablespoon iced water

CHEESE FILLING
80 g (2¾ oz) cream cheese, at room
    temperature
20 g (¾ oz) butter, softened
1 tablespoon finely snipped chives
1 tablespoon finely chopped flat-leaf
    (Italian) parsley
½ teaspoon finely grated lemon zest
90 g (3¼ oz/¾ cup) finely grated cheddar
    cheese

Preheat the oven to 190°C (375°F/Gas 5). Line two baking trays with baking paper.

To make the biscuit pastry, sift the flour and baking powder into a large bowl then add the chopped butter. Using your fingertips, rub in the butter until the mixture resembles fine breadcrumbs. Make a well in the centre, add the egg, cheese, herbs and iced water to the well and mix with a flat-bladed knife until a coarse dough forms. Transfer the mixture to a lightly floured surface and gather together into a ball.

Roll the pastry out between two sheets of baking paper until 3 mm (⅛ inch) thick. Remove the top sheet and cut the pastry into rounds, using a 5 cm (2 inch) cutter. Place the rounds on the baking trays. Re-roll pastry scraps and cut out more rounds. Bake the biscuits for 8 minutes, or until light brown. Transfer to a wire rack to cool.

To make the filling, beat the cream cheese and butter in a small bowl using electric beaters until light and creamy. Add the herbs, lemon zest and cheese, season to taste with freshly ground black pepper and beat until smooth. Spread ½ teaspoon of the filling on half the biscuits and sandwich with the remaining biscuits.

Unfilled biscuits will keep, stored in an airtight container in a cool, dark place, for 2 days. Filled biscuits are best served on day of filling.

SPINACH AND LEEK FRITTERS
MAKES 8

40 g (1^1/$_2$ oz) butter
40 g (1^1/$_2$ oz/^{1}/$_4$ cup) pine nuts
1 leek, white part only, thinly sliced
100 g (3^1/$_2$ oz/2 cups) English spinach
  leaves, chopped
3 eggs
1 egg yolk
1 tablespoon pouring (whipping) cream
75 g (2^1/$_2$ oz/^{3}/$_4$ cup) grated parmesan
  cheese
1 tablespoon chopped flat-leaf (Italian)
  parsley
1 tablespoon olive oil

Melt half the butter in a heavy-based frying pan over medium-low heat, add the pine nuts and leek and cook, stirring often, for 3 minutes, or until the pine nuts are golden. Add the spinach and cook for 1 minute. Remove the mixture from the pan and cool slightly. Wipe the pan with paper towels.

Whisk the eggs, egg yolk and cream together in a large bowl. Add the cheese and parsley and season with salt and freshly ground black pepper. Stir in the spinach mixture.

Melt half of the remaining butter and half of the oil in the frying pan. Place four greased 5–7 cm (2–2^3/$_4$ inch) egg rings in the pan, and pour 60 ml (^{1}/$_4$ cup) of the spinach mixture into each. Cook over low heat for 2–3 minutes, or until the base is set. Using a spatula, carefully turn the fritters in the egg rings and cook the other side for 2–3 minutes, or until firm. Transfer to a plate and remove from the egg rings. Cover the plate loosely with foil and place in a low oven to keep warm while you cook the remaining fritters. Repeat with the remaining butter, oil and spinach mixture to make 8 fritters in total. Serve immediately.

POTATO, GRANA PADANO, OLIVE AND ROSEMARY CAKE
SERVES 4–6

8 all-purpose potatoes (such as desiree)

30 g (1 oz) butter

2 tablespoons olive oil

1 garlic clove, crushed

200 g (7 oz/2 cups) dried breadcrumbs

1½ tablespoons rosemary leaves, finely chopped

120 g (4¼ oz/¾ cup) pitted black olives, finely chopped

100 g (3½ oz) grated cheddar cheese

75 g (2½ oz/¾ cup) freshly grated grana padano cheese

Preheat the oven to 180°C (350°F/Gas 4). Grease a deep 20 cm (8 inch) springform cake tin. Line the base and side with baking paper.

Thinly slice the potatoes. Heat the butter and oil in a small frying pan, add the garlic and ½ teaspoon freshly ground black pepper and remove from the heat.

Place the potato slices in the base of the tin, overlapping them to neatly cover the base. Brush the potatoes with some of the butter mixture. Combine the breadcrumbs, rosemary, olives and cheese in a bowl, then sprinkle some of the mixture over the potatoes in the tin. Continue layering until the potatoes, butter mixture and cheese mixture are used up, then press down firmly. Bake for 1 hour or until golden and crisp, then serve immediately, cut into wedges.

Veal parmigiana
Serves 4

4 thin veal steaks
100 g (3½ oz/1 cup) dry breadcrumbs
½ teaspoon dried basil
25 g (¾ oz/¼ cup) finely grated parmesan
  cheese, plus 50 g (1¾ oz/½ cup) extra
plain (all-purpose) flour, for coating
1 egg, lightly beaten
1 tablespoon milk
olive oil, for frying
250 g (9 oz/1 cup) tomato passata (puréed
  tomatoes)
100 g (3½ oz/⅔ cup) grated mozzarella
  cheese

Trim the meat of any excess fat and sinew, place between two sheets of plastic wrap and gently pound with a meat mallet until 5 mm (¼ inch) thick. Snip the edges to prevent the meat from curling while cooking.

Combine the breadcrumbs, basil and parmesan on a sheet of baking paper.

Dust the veal steaks in flour, shaking off any excess. Working with one at a time, dip the steaks into the combined egg and milk, allowing any excess to drip off. Coat with the breadcrumb mixture, lightly shaking off the excess. Refrigerate for 30 minutes to firm the coating.

Preheat the oven to 180°C (350°F/ Gas 4). Heat the oil in a frying pan and brown the veal steaks over medium heat for 2 minutes on each side, working in batches if necessary. Drain on paper towels.

Spread half the tomato passata into a shallow ovenproof dish. Arrange the veal steaks on top in a single layer and spoon over the remaining sauce. Top with parmesan cheese and mozzarella and bake for 20 minutes, or until the cheeses are melted and golden brown. Serve immediately.

FRENCH ONION SOUP
SERVES 4

50 g (1¾ oz) butter
1 tablespoon olive oil
1 kg (2 lb 4 oz) onions, thinly sliced into
  rings
3 x 425 g (15 oz) tins chicken or beef
  consommé
125 ml (4 fl oz/½ cup) dry sherry
half a baguette
125g (4½ oz/1 cup) finely grated cheddar
  or gruyère cheese

Heat the butter and oil in a large saucepan, add the onion and cook, stirring frequently, over low heat for 45 minutes, or until softened and translucent. It is important not to rush this stage — cook the onion thoroughly so that it caramelises and the flavour develops.

Add the consommé, sherry and 1 cup (9 fl oz/250 ml) water. Bring to the boil then reduce the heat and simmer for 30 minutes. Season to taste.

Meanwhile, slice the bread into four thick slices and arrange them in a single layer under a hot grill. Toast one side, remove from the grill, turn over and cover the untoasted side with the gruyere.

Ladle the hot soup into four serving bowls, top each with a slice of toast, cheese side up, and place under the grill until the cheese is melted and golden.

FONDUE
SERVES 6

1 garlic clove, halved
350 ml (12 fl oz) dry white wine
1 tablespoon lemon juice
400 g (14 oz) emmental cheese, rind
  removed, cut into small pieces
400 g (14 oz) gruyere cheese, rind
  removed, cut into small pieces
80 ml (2½ fl oz/⅓ cup) kirsch
1 tablespoon cornflour (cornstarch)
a large pinch of freshly grated nutmeg
sea salt
2 baguettes, cut into 2.5 cm (1 inch)
  pieces

Rub the inside of the fondue pot with the garlic, then discard the garlic. Heat 300 ml (10½ fl oz) of the wine in the pot until quite hot but not simmering, then add the lemon juice. Add the cheese, half a cup at a time, stirring constantly over medium-low heat until the cheese has melted and the mixture is smooth. Add the remaining wine and stir to combine well.

In a small bowl, combine the Kirsch and cornflour and stir until a smooth paste forms. Add to the fondue, then cook the mixture for 1–2 minutes, or until it thickens slightly. Add the nutmeg, season to taste with sea salt and freshly ground black pepper, then serve, placed over a burner at low heat, with bread passed separately.

index

Published in 2008 by Murdoch Books Pty Limited

Murdoch Books Australia
Pier 8/9
23 Hickson Road
Millers Point NSW 2000
Phone: +61 (0) 2 8220 2000
Fax: +61 (0) 2 8220 2558
www.murdochbooks.com.au

Murdoch Books UK Limited
Erico House
6th Floor
93–99 Upper Richmond Road
Putney, London SW15 2TG
Phone: +44 (0) 20 8785 5995
Fax: +44 (0) 20 8785 5985

Chief Executive: Juliet Rogers
Publishing Director: Kay Scarlett

Design Manager: Vivien Valk
Project Manager and Editor: Janine Flew
Design concept: Sarah Odgers
Design: Alex Frampton
Production: Monique Layt
Photographer: George Seper
Stylist: Marie-Hélène Clauzon
Food preparation: Joanne Glynn

National Library of Australia Cataloguing-in-Publication Data
Kitchen, Leanne
Title: The dairy / author, Leanne Kitchen.
ISBN: 9781741962017 (hbk.)
Series: Providore series
Notes: Includes index.
Subjects: Dairy products.
Other Authors/Contributors: Flew, Janine.
Dewey Number: 637

Colour reproduction by Splitting Image Colour Studio, Melbourne, Australia.
Printed by 1010 Printing International Limited in 2008. PRINTED IN CHINA.

CONVERSION GUIDE: You may find cooking times vary depending on the oven you
are using. For fan-forced ovens, as a general rule, set the oven temperature to 20°C (35°F)
lower than indicated in the recipe.